Compressed Hydrogen in Fuel Cell Vehicles

Compressed Hydrogen in Fuel Cell Vehicles

On-board Storage and Refueling Analysis

Shitanshu Sapre, Kapil Pareek
and Rupesh Rohan

CRC Press
Taylor & Francis Group
Boca Raton London New York

CRC Press is an imprint of the
Taylor & Francis Group, an **informa** business

First edition published 2022
by CRC Press
6000 Broken Sound Parkway NW, Suite 300, Boca Raton, FL 33487-2742

and by CRC Press
2 Park Square, Milton Park, Abingdon, Oxon, OX14 4RN

CRC Press is an imprint of Taylor & Francis Group, LLC

ISBN: 978-1-032-15489-3 (hbk)
ISBN: 978-1-032-15490-9 (pbk)
ISBN: 978-1-003-24431-8 (ebk)

DOI: 10.1201/9781003244318

Typeset in Times LT Std
by SPi Technologies India Pvt Ltd (Straive)

Contents

Preface

With a share of 21% of total energy consumption and 23% of emission greenhouse gases worldwide, transport has become the second largest contributor toward present energy crises. In order to deal with this, the development of clean transportation with revolutionary automotive technologies such as battery-operated electric vehicles, hybrid vehicles and hydrogen fuel cell vehicles is required. Particularly, hydrogen fuel cell vehicle has many benefits such as high energy conversion, efficient drivetrain and zero carbon dioxide emission than conventional gasoline vehicles. However, the efficient onboard storage of hydrogen to achieve the desired drive range is a significant challenge for the market acceptability of hydrogen fuel cell vehicles. This efficient refueling of compressed hydrogen in vehicle tanks plays an important role in achieving desired storage density leading to the required drive range.

The book presents a detailed investigation of refueling of compressed hydrogen in Type IV composite tanks used in fuel cell vehicles. It includes the assessment of refueling of compressed hydrogen using computational fluid dynamic simulation. The book introduces the constant and variable refueling approach for refueling compressed hydrogen. The role of heat transfer in refueling has been investigated by the heat capacity model. It also covers the influence of severe refueling conditions on the structure of the composite tank.

The content of the book describes the work that had been carried out during the Ph.D. study of Dr. Shitanshu Sapre in the Energy Storage Lab at Centre for Energy and Environment of Malaviya National Institute of Technology Jaipur led by Dr. Kapil Pareek. The various chapters or sections are based on scientific papers published in various journals.

Author Biographies

Dr. Shitanshu Sapre worked as Research Fellow in Energy Storage Lab of Centre for Energy and Environment in Malaviya National Institute of Technology Jaipur. He is also affiliated with Bhartiya Skill Development University, Jaipur. His research area is hydrogen storage, hydrogen infrastructure and fuel cell vehicle.

Dr. Kapil Pareek working as Assistant Professor at the Centre for Energy and Environment in Malaviya National Institute of Technology, Jaipur. He is also the leader of the Energy Storage Lab where various research scholars work on Energy storage technologies such as hydrogen storage, batteries and fuel cells. His expertise includes hydrogen storage technologies, fuel cells, battery storage, thermal management of battery and battery stack. He has completed his Ph.D. from National University of Singapore, Singapore, in 2014 with a focus on hydrogen storage technology.

Dr. Rupesh Rohan works under the capacity of Assistant Director & Centre Head of Indian Rubber Manufacturer Research Association, South Centre, Sricity (AP). He is a material scientist and completed his Ph.D. from National University of Singapore, Singapore, in 2015 with a focus on energy storage technology.

Abbreviations

ASME	American Society of Mechanical Engineers
ANSI	American National Standards Institute
CH$_2$	Compressed hydrogen
CCH$_2$	Cryo-compressed hydrogen
Capex	Capital expenditure
CFRP	Carbon fiber reinforced polymer
CGA	Compressed gas association
CEN	European Committee for Standardization
CFD	Computational fluid dynamics
EOS	Equation of state
FCV	Fuel cell vehicle
FW	Filament winding
GasTef	Gas testing facility
HDPE	High density polyethylene
HTC	Heat transfer coefficient
HCM	Heat capacity model
ISO	International Organization for Standardization
JRC	Joint research centre
LH$_2$	Liquid hydrogen
LLNL	Lawrence Livermore National Laboratory
LOHC	Liquid organic hydrogen carriers
MOF	Metal–organic framework
NFPA	National Fire Protection Association
PA-6	Polyamide 6
PET	Polyethylene terephthalate
RCA	Root cause analysis
RMSE	Root means square error
SAE	Society of Automotive Engineers
SOC	State of charge
US DOE	United States Department of Energy

1 Introduction
Hydrogen Storage Techniques

1.1 HYDROGEN STORAGE METHODS

For the development of the future hydrogen economy, a safe and efficient means of storing hydrogen is required in onboard, portable and stationary applications. Storage is a challenging issue that cuts across production, delivery and end-use applications of hydrogen as an energy carrier. Indeed, for a successful application of hydrogen as an energy carrier, hydrogen should be stored safely and efficiently as conventional fuels [1]. One of the most challenging applications in this field is hydrogen storage for mobile or onboard applications. Hydrogen storage is a key enabling technology for the successful penetration of hydrogen fuel cell vehicles in the gasoline vehicle market [2].

The size, weight, and density are deciding factors for any storage system for particular applications. However, sometimes characteristics of fuel also play a significant role in defining the suitability of the storage system. Hydrogen possesses unusual physical and thermodynamic properties that make it difficult to call any storage system as efficient for particular applications. Under normal temperature and pressure, the density of the gas is very low near about 0.08238 kg/m^3, e.g. for storing 5 kg of hydrogen, which implies a volume of around 60 m^3 and energy content of 600 MJ (166.65 kWh). For the same weight and energy content, the gasoline volume is 0.019 m^3.

Given these numbers, it is clear that for efficient storage, hydrogen density should be increased by reducing the volume taken by the gas under normal temperature and pressure conditions. As a consequence, the "normal state" of hydrogen has to be changed in order to store it efficiently. This can be accomplished by increasing the pressure, decreasing the temperature below the critical temperature or reducing the repulsion interaction between hydrogen molecules by binding them with another material.

Thus, for storing hydrogen, various methods have emerged as a perfect solution for hydrogen storage. They have been classified as (1) physical storage (2) materials-based storage (as shown in Figure 1.1). However, all methods have their advantages and disadvantages based on their storage capacity and operating conditions.

DOI: 10.1201/9781003244318-1

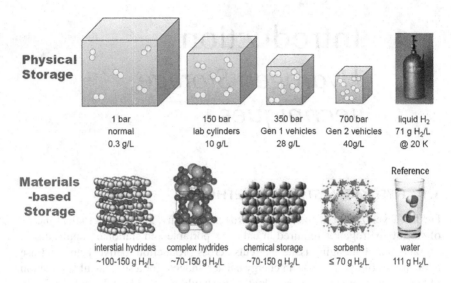

Physical Storage

1 bar	150 bar	350 bar	700 bar	liquid H$_2$
normal	lab cylinders	Gen 1 vehicles	Gen 2 vehicles	71 g H$_2$/L
0.3 g/L	10 g/L	28 g/L	40g/L	@ 20 K

Materials -based Storage

Reference

interstial hydrides	complex hydrides	chemical storage	sorbents	water
~100-150 g H$_2$/L	~70-150 g H$_2$/L	~70-150 g H$_2$/L	≤ 70 g H$_2$/L	111 g H$_2$/L

FIGURE 1.1   Various hydrogen storage methods [3].

1.2   PHYSICAL HYDROGEN STORAGE

In this class, hydrogen at normal state is converted into high pressure or low temperature gas. Based on the storage pressure and temperature, the storage system is subclassified as liquid, cryo-compressed and compressed. All these approaches are used to increase the storage density of hydrogen (Table 1.1).

For hydrogen fuel cell passenger cars, around 5–6 kg of hydrogen storage with high gravimetric and volumetric storage density is required to achieve the drive range of more than 500 km. For this, pure hydrogen can be stored as liquid at −253°C or high-pressure gas up to 700 bar in a suitable container.

1.2.1   LIQUID HYDROGEN STORAGE

Hydrogen density can be increased by liquefying the hydrogen to attain a storage density of 70 g/L at 1 bar and −253°C. Liquid hydrogen is mainly considered as a distribution method to take advantage of high hydrogen density at low temperature. During the liquefaction of hydrogen, the ortho hydrogen is converted into

TABLE 1.1
Physical Hydrogen Storage Methods [5,6]

Hydrogen storage methods	Storage Density (g/L)	Operating Pressure and Temperature
Compressed	40	70 MPa, 15°C
Liquid	71	0.1 MPa, −253°C
Cryo-compressed	90	35 MPa, −233°C

the para form with heat release that evaporates the liquefied hydrogen into the gaseous one backward. The ortho-para conversion catalysts are usually used during liquefaction to avoid such boil-off.

Liquid hydrogen occupies less volume than pressurized storage and is particularly attractive for attaining higher storage densities. However, some unusual properties of liquid hydrogen such as low boiling point, negative Joule–Thomson Coefficient and low vaporization enthalpy impose technical barriers to its acceptability. The low boiling point of hydrogen necessitates a high level of purification of less than 1 ppm to avoid clogging. Similarly, low vaporization enthalpy generates the need of well-insulated storage vessels.

The cryogenic liquid starts to evaporate after a certain period of time called the boil-off phenomenon. This leads to the 2–3% loss of hydrogen per day and extra energy input is required for storage vessels. This cannot be prevented, even with a very effective vacuum insulation and heat-radiation shield in place. Hydrogen boil-off is considered an issue in terms of refueling frequency, cost, energy efficiency and safety, particularly for vehicles parked in confined spaces, such as parking garages [7]. Therefore, each technology requires well-insulated and expensive cryogenic storage vessels to prevent boil-off and maximize dormancy. Figure 1.2 shows generation 2 cryogenic pressure vessel with aluminum-lined, carbon fiber wrapped surrounded by a vacuum space for storing cryogenic hydrogen [8].

Moreover, the liquefaction process is energy-intensive and consumes approximately 35% of the energy content of the stored hydrogen. Therefore, liquid hydrogen is limited to flight and space applications where high volumetric and

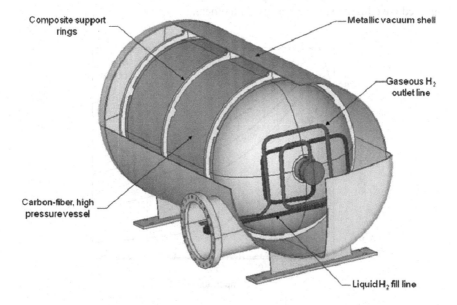

FIGURE 1.2 Generation 2 cryogenic capable pressure vessel design [7].

gravimetric energy storage densities are required regardless of its high-power consumption. The net energy consumed is approximately 10 kWh/kg and it contributes 40–50% of capital expenditure (CapEex) of the liquid hydrogen storage system [9]. Hence, energy expenditure and boil-off due to leakage is a major challenge to the liquid hydrogen storage for long-term storage and of hydrogen storage for automotive applications.

1.2.2 Cryo-compressed Hydrogen Storage

Cryo-compressed, currently the latest approach for storing the hydrogen at a pressure above normal state and temperature similar to or less than liquid hydrogen. It is a combination of two approaches where hydrogen is compressed to higher pressure up to 35 MPa and temperature −233°C. The high pressure and cryogenic storage increase the gravimetric and volumetric density while overcoming the boil-off loss (dormancy) from liquefied hydrogen storage. As a result of increasing the pressure of liquefied hydrogen the storage density increases to 90 g/L [10].

The density of cryo-compressed is much higher than compressed and relatively increased compared to the liquid hydrogen as in Figure 1.3 due to the increase in pressure of storage. Hydrogen can be stored in an insulated tank at a temperature of 20 K and pressure up to 35 MPa. This type of cryogenic tank can withstand the high pressure and significantly extend the time before starting evaporative losses when they are in operation and thus increase storage autonomy.

The BMW has demonstrated the cryo-compressed storage for onboard application in their FCV named BMW 7. Figure 1.4 presents the comparison done by BMW on all three physical storage systems which reported that the cryo-compressed have higher storage density than others.

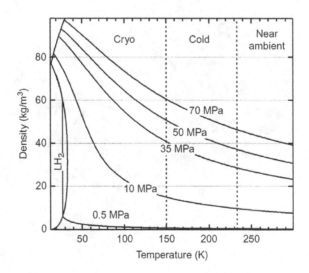

FIGURE 1.3   Hydrogen density at temperature and pressure [12].

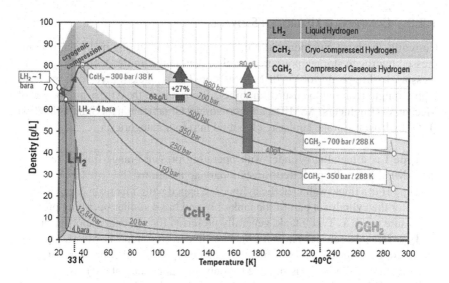

FIGURE 1.4 Hydrogen density at different pressures and temperatures by BMW [11].

The Lawrence Livermore National Laboratory (LLNL) has demonstrated and tested the generation 2 type III composite pressure vessel with a metallic liner that is encapsulated in a secondary insulated jacket, whose role is to limit heat transfer between the hydrogen and the environment (as shown in Figure 1.5) [12]. The most significant advantage of cryo-compressed approach is the flexibility in fuel as a vessel is designed to operate at moderate pressure to higher pressure. Additionally, it also extends the dormancy period and allows the gaseous phase as the temperature of gas increases with heat transfer.

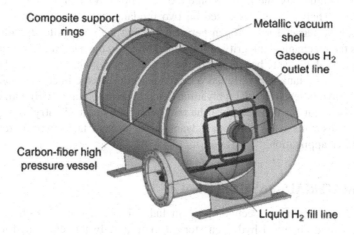

FIGURE 1.5 Type III cryo-compressed hydrogen tank by LLNL [12].

1.2.3 COMPRESSED HYDROGEN STORAGE

Currently, the most common and efficient approach of storing hydrogen is compressed hydrogen storage pressure vessels called tanks. In this technology, high-pressure hydrogen up to 35–70 MPa is stored in a thick-walled tank made of high-strength material to ensure durability. According to the American Society of Mechanical Engineers (ASME) and International Organization for Standardization (ISO) hydrogen storage tanks are classified into four categories:

Type I is a metallic cylinder with operating pressure range limited to 20 MPa, generally preferred for stationary applications.

Type II is a thick-wall metallic cylinder wrapped with fiber resin composite for the cylindrical part of the tank only and has a similar operating pressure range as Type I tank, generally preferred for stationary application.

Type III tanks consist of metallic liner and fiber resin composite overwrapped along the surface of the cylinder. It can be used for higher pressure range up to 45 MPa which is suitable for onboard application of the tank.

Type IV is a polymer-lined tank overwrapped with fiber resin composite. Due to the high strength composite wrapping the operating pressure range increases to more than 100 MPa. Presently, Type IV tanks are preferred for onboard application.

Type V is a liner-less tank with a significant reduction in weight and higher storage capacity compared to Type III and IV. The tank has a better weight to tank capacity ratio and is currently demonstrated for the aircraft/aerospace industry.

The most advanced lightweight storage system for onboard application is a composite tank with non-load carrying metallic (Type III) or polymer (Type IV) liner composite tanks. Type III and IV tanks have attracted much interest from the scientific community and also become the first choice for fuel cell vehicle manufacturers. Their structure is based on two fundamental components: the liner, essentially a barrier for hydrogen permeation, and the composite structure that ensures the mechanical integrity of the tank. However, the metallic embrittlement of liner at high pressure has restricted the use of Type III tanks nowadays [13].

The Type IV tank generally has a combination of plastic liner made of high-density polyethylene (HDPE), polyamide, etc., and carbon fiber (CF) with epoxy resin which reduces the weight of the tank with a higher load-carrying capacity [14,15]. This leads to the additional advantage for Type IV tanks to be used in fuel cell vehicle application.

1.3 MATERIALS-BASED STORAGE

Material-based storage technologies include metal hydrides, sorbent-based materials, and chemical hydrogen storage materials. In this class, hydrogen is reversibly absorbed or adsorbed by two mechanisms known as physisorption and chemisorption. In physisorption, hydrogen molecules are absorbed in the surface of the material by weak Van der Waals forces. However, chemisorption involves the dissociation of hydrogen molecules and the chemical bonding of atoms with the host matrix. Hydrogen stored in solid-state form has the potential advantage of higher volumetric density compared to physical storage methods.

1.3.1 METAL HYDRIDES

Metal hydrides are currently promising options for many stationary and mobile hydrogen storage applications. Hydrides are formed by transition metals including rare earth metals. Metallic hydrides have a wide variety of stoichiometric and non-stoichiometric compounds and are formed by direct reaction of hydrogen with the metal or by electrochemical reaction as shown in Figure 1.6 [12,16]. Interstitial hydrides are formed from metallic elements or compounds that react with gaseous hydrogen to produce binary, or higher, hydrides, intermetallic compounds, solid solution alloys, amorphous and nanostructured hydrides.

Hydrogen-absorbing intermetallics form a number of different groups, which can be distinguished by their stoichiometries, including AB5, A2B7, AB3, AB2, AB and A2B compounds. The archetypal AB5 intermetallic is LaNi5. This compound readily forms a hydride under fairly moderate hydrogen pressures and ambient temperatures. AB2 compounds can be formed from the combination of many different elements where A can be from group 4 (Ti, Zr, Hf) or the lanthanoids (La, Ce, Pr, and so forth), whereas the B element can be a transition or non-transition metal, with a preference for V, Cr, Mn and Fe. In comparison to the number of different AB2 and AB5 compositions reported in the literature, the number of AB compounds of interest for hydrogen storage is fairly limited. Solid solution alloys formed by dissolving one or more hydrogen-absorbing metallic

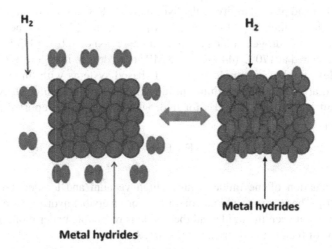

FIGURE 1.6 Schematic of hydrogen storage in metallic hydrides [16].

elements in another, for instance, Ti–V–Fe (Ti43.5V49.0Fe7.5) and Ti–V–Cr–Mn alloys exhibit typical gravimetric hydrogen capacities up to 4 wt% [18].

MgH_2 is the binary hydride that has attracted by far the most attention as a potential storage material; however, there are two other binary hydrides that we should also mention, namely AlH_3 and PdHx. The former is of interest due to its high gravimetric storage capacity of 10.1 wt% [19–21]. However, it is effectively non-reversible within a realistic hydrogen pressure range for a practical storage unit and therefore requires off-board regeneration. This process is economically and energetically costly.

Apart from the interstitial hydrides, complex hydrides have attracted great attention for hydrogen storage applications. The complex hydrides are formed via the reaction of atomic hydrogen either ionically or covalently into the bulk of alkali or alkaline earth metals and the hydrogen is released via the decomposition of the host into two or more components. The term complex hydride has become an umbrella term encompassing the alanates, nitrides and borohydrides that are currently being considered for hydrogen storage [22,23]. A well-known example is sodium alanate ($NaAlH_4$) [24]. During the dehydrogenation process, this phase decomposes into an intermediate Na_3AlH_6 phase with an associated release of gaseous hydrogen around 493 K and a second step around 525 K.

$$3NaAlH_4 \quad \rightarrow \quad Na_3AlH_6 + 2Al + 3H_2$$

A second reaction step then results in further hydrogen evolution,

$$Na_3AlH_6 \rightarrow 3NaH + Al + 3/2\,H_2$$

The dehydrogenation of NaH does not occur until 698 K and the above two-stage decomposition shows two plateaus in the isotherm. The irreversibility of this reaction and the instability of the hydride with slow desorption kinetics were addressed by adding $TiCl_3$ to improve its performance [24]. Ti-doped $NaAlH_4$ readily desorbs hydrogen at temperatures in the region of 120°C (393 K) and can be hydrogenated at 170°C (443 K) in 15 MPa of hydrogen in Figure 1.7 [25].

Nitrides, amides and imides (Li–N–H)-based systems with three stoichiometric ternary compounds: lithium imide (Li_2NH), lithium amide ($LiNH_2$) and lithium nitride hydride (Li_4NH) for hydrogen storage were reported by Chen et al. [26].

$$Li_3N + 2H_2 \quad \rightarrow \quad Li_2NH + LiH + H_2 \leftrightarrow LiNH_2 + 2LiH$$

Dehydrogenation of the imide requires high vacuum and temperatures above 600 K [26]. These are unsuitable conditions for reversible hydrogen storage, but the reaction between the imide and the amide is reversible under more moderate conditions of both temperature and pressure. A number of other similar materials have been studied as hydrogen storage media, including the ternary compounds $Mg(NH_2)_2$, $RbNH_2$, $CsNH_2$ and Ca–N–H, and the quarternary and higher systems

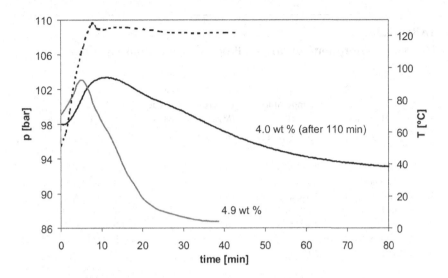

FIGURE 1.7   Hydrogenation curves of Ti-doped NaAlH₄ [12].

Li–Ca–N–H, Li–Al–N–H, Na–Mg–N–H, Na–Ca–N–H, Mg–Ca–N–H and Li–Mg–Ca–N–H [23]. It can be seen that these materials have the potential to provide high gravimetric storage capacities and are therefore of great interest. However, the Li–N–H system suffers from some drawbacks, including high hydrogenation and dehydrogenation temperatures, and air or moisture sensitivity. Another is the evolution of ammonia during the dehydrogenation reaction [27].

Borohydrides (LiBH₄) have the highest gravimetric hydrogen storage capacities up to 18.5 wt% hydrogen but suffer from very high decomposition temperature [23]. LiBH₄ releases three of its four hydrogen atoms upon melting at 553 K, with an enthalpy of decomposition of −88.7 kJ/mol H₂. LiH is very stable and its dehydrogenation occurs only above a temperature of 1000 K. Nevertheless, the dehydrogenation reaction of LiBH₄ is a reversible process, although rehydrogenation requires elevated pressures and temperatures [23].

$$LiBH_4 \rightarrow Li + B + 2H_2$$

$$LiBH_4 \rightarrow LiH + B + 3/2\,H_2$$

A number of other alkali metal and alkaline earth metal borohydrides have high gravimetric and volumetric hydrogen storage capacities. For example, the theoretical gravimetric capacities of NaBH₄, KBH₄ and Mg(BH₄)₂ are 10.6, 7.4 and 14.8 wt%, with volumetric capacities of 113.1, 87.1 and 146.5 kg/m³, respectively. However, these are all theoretical values and cannot be reversibly achieved in practice at practical temperatures; borohydrides are also moisture sensitive [28] and Eberle et al. [29] suggest that the possible evolution of volatile boranes, even at trace levels, would be problematic due to storage capacity loss and fuel cell damage.

TABLE 1.2

Hydrogen Absorption/Desorption Properties of Various Hydrides

S. No	Hydrides	Temperature (K)	Pressure (MPa)	Hydrogen Storage Capacity (wt%)	Reference
1	$CeNi_4Zr$	Tdes: 293–333	3.2	4	J Alloys Compd 2007;440(1e2):84–8
2	$LiBH_4$ nanocomposite	Tdes: 573	0.1	9.6	Dalton Trans 2008;(40):5400–13
3	$NaAlH_4$ + 1.0 mol% Ti	Tabs: 443 Tdes: 423	15.4	5.6	Appl Phys A 2001;72(2):213–9.
4	NaAlH4 + Porous carbon	Tdes: 673	10	7	J Phys Chem C 2010;114(10):4675–82.
5	$LaNi_5H_6$	295	–	1.37	J Alloys Compd 2007;436(1e2):345–50
6	$Ti_{0.5}V_{0.5}Mn$	Tdes: 260	35	1.90	J Alloys Compd 2013;580(Supplement 1(0)):S233–7.
7	$CeNi_3Cr_2$	Tdes: 293–333	3.25	3.80	J Alloys Compd 2007;430(1e2):165–9.
8	$NaAlH_4$ + Nonporous carbon	Tdes: 673	10	6.3	J Phys Chem C 2010;114(10):4675–82.

A list of some interesting metal hydrides and their hydrogen sorption properties are tabulated in Table 1.2.

1.3.2 SORBENTS

Sorbents are based on weak Van der Waals bonds between molecular hydrogen and material molecules. In the physisorption, the hydrogen is absorbed on the surface of the solid so materials with high specific surface area are required for molecular hydrogen storage. Compared to complex hydrides, restructuring and formation of chemical bonds do not take place during storage. Therefore, sorbents have a long-lifetime during cycling [30].

The advantages of sorbents mainly include the absence of high-pressure requirement as in compressed storage, do not need high energy input as in cryo-compressed and liquid hydrogen demands, and higher temperature desorption as in complex hydrides [30]. The widely used materials in this category are carbon-based materials, metal–organic frameworks (MOFs), zeolites, microporous metal coordination materials, etc.

Carbon-based materials including activated carbons, carbon nanotubes, nanofibers and microporous templated carbons [31] have been utilized for hydrogen storage application via physical adsorption. Carbon is very attractive technically as a host because of its low molar mass. It is also chemically stable and can be synthesized in a number of different forms. From a practical point of view, porous carbons are already commercially produced in large quantities for a broad range of applications and are relatively inexpensive. The activated carbons, as reported by Yürüm et al. [31], can show that gravimetric storage capacities can reach 5.5 wt% at 77 K. The hydrogen sorption properties of carbon nanostructures, such as nanotubes and nanofibers, have been investigated extensively in recent years. Surface area, pore size and pore size distribution play an important role during hydrogen storage of sorbent materials. Some of the carbon-based materials and their hydrogen storage properties are tabulated in Table 1.3.

MOFs are a class of inorganic–organic hybrid porous crystalline materials consisting of metal ions. MOFs have several properties that make them particularly attractive for hydrogen storage such as their extraordinarily high surface areas, ultrahigh porosities, and modifiable internal surfaces. MOF has gained attention for research by many researchers as they have good stability, large surface area, and adjustable pore size [39]. The first MOF, MOF-5 was synthesized in 1999 by the group of Omar Yaghi and his collaborators from the University of California at Los Angeles. This MOF which is inexpensive to manufacture, lightweight and stable, is already able to store a quantity of hydrogen up to 4.5% by weight at 77 K and 0.7 bar [40].

Apart from this, some of the MOFs which have been used for efficient hydrogen storage along with their storing capacities are isoreticular metal–organic frameworks (IRMOFs) such as IRMOF-1 (1.3 wt%), MOF-177 (7.5 wt%), IRMOF-20 (6.5%) and Li-doped MOF-C-30 (6 wt%) at −30.1C wt% at 50 bar) [41–43]. Interaction of hydrogen with MOF is shown in Figure 1.8 with three orientations of the adsorbed hydrogen molecule (horizontal, vertical and sloping) in each adsorption site [44].

TABLE 1.3
Hydrogen Storage in Carbon-based Sorbent Materials

Material	Temperature	Hydrogen Storage Capacity (wt%)	Reference
PAN-derived ACs	77	0.58–1.95	[32]
Coconut shell-derived ACs	77	1.1–2.15	[32]
CNT	298 K and 10 MPa	>10	[33,34]
Porous carbon	77 K and 0.1 MPa	0.5–7.5	[35]
Pretreated SWCNTs	298 K and 10 MPa	2.4–4.2	[36]
Graphitic carbons	100–200 K and 0.5 MPa	0.09–1.1	[37]
Activated carbons	298 K and 10 MPa	1	
Heat-treated MWCNTs	298 K and 10 MPa	1.3–4.0	[38]
SWCNTs	298 K and 10 MPa	4–5	[36]

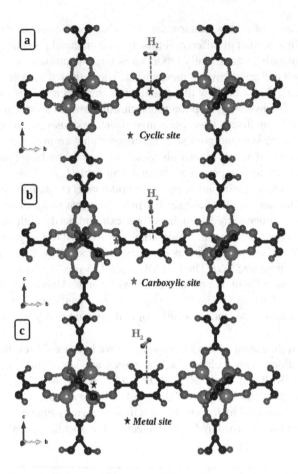

FIGURE 1.8 H_2 molecule adsorption in three different adsorption sites and along three orientations: (a) horizontal orientation, (b) vertical orientation and (c) sloping orientation [43].

Zeolites are crystalline alumino-silicate materials having highly ordered structures with good adsorbents for various gases such as hydrogen carbon dioxide as well as moisture. The absorption of hydrogen on zeolites depends on the surface area of adsorption [17]. Hydrogen is adsorbed at the surface of zeolites forcing into a porous structure at high pressure and temperature. The hydrogen storage capacity of zeolites can improve doping by some other materials like palladium and similar metals [45]. Zeolites are known for their high thermal stability and low cost. Some of the zeolites and their hydrogen storage properties at 77 K are tabulated below.

1.3.3   LIQUID ORGANIC HYDROGEN CARRIERS

Liquid organic hydrogen carriers (LOHCs) are organic compounds used for storing hydrogen via a chemical reaction. LOHCs are unsaturated compounds

TABLE 1.4

Approximate Pore Size of Zeolites and Hydrogen Storage Properties of Zeolites at 77 K [46–53]

Framework type	Framework Type (IZA Code)	Hydrogen Uptake wt%	Pressure MPa	Channel Size (nm)	Surface Area (m³/g)	Pore Volume (mL/g)
NaA	LTA	1.54	1.5	0.4	–	0.3
NaX	FAU	1.79	1.5	0.74	662	0.36
NaY	FAU	1.81	1.5	0.74	725	0.34
Sodalite	SOD	~2.75	1	0.28	–	0.5
H-Chabazite (Si/Al = 2.125)	CHA	1.1	0.092	0.43	490	–
ZSM-5	MFI	0.71	0.1	0.53 × 0.56	431	0.28
ZSM-35	FER	0.58	0.1	054 × 0.42	344	0.32
Zeolite L	LTL	0.53	0.1	0.8–0.9	344	0.25
H-MOR (Si/Al = 7.0)	MOR	0.6	0.66	0.7 × 0.65	–	0.32
MCM-41	–	1.6	3.5	2.2 ± 0.2	1017	1.04

with double or triple carbon that take the hydrogen during hydrogenation. The hydrogenation is an exothermic reaction carried out at pressure and temperature of 30–50 bar and 150–200°C in the presence of a catalyst, respectively [54,55]. The storage of hydrogen takes place by reversible hydrogenation and dehydrogenation of carbon double bonds. Hydrogenation is an exothermic process at high pressure and temperature and dehydrogenation is an endothermic process at atmospheric pressure. Hydrogenation and dehydrogenation are catalyst-based processes where catalysts play a significant role in conducting the reaction. Figure 1.9 presents the hydrogenation and dihydrogen process of hydrogen in typical LOHC compounds with heat generation and absorption during the process [56].

The gravimetric storage density of unsaturated organic compounds is about 6% by weight. Due to the molecular bonding of hydrogen in LOHC materials, the volumetric density of LOHC is greatly increased leading to easier transportation similar to crude oil. It can be used with the existing infrastructure of crude oil transport. Apart from this, higher volumetric density LOHC materials are cheap and easily available. They generally have non-toxic properties and low dehydrogenation temperature. However, LOHC requires higher energy during hydrogenation and dehydrogenation [55].

N-ethyl carbazole (NEC) is a commonly used LOHC having nitrogen-substituted heterocycle. The hydrogenated form is called perhydro-N-ethyl carbazole. Figure 1.10 shows hydrogenation and dehydrogenation process of N-ethyl carbazole. Hydrogen and dehydrogenation take place in the presence of palladium (Pd) and ruthenium (Ru) catalyst resulting in a storage density of 5.8% by weight [56].

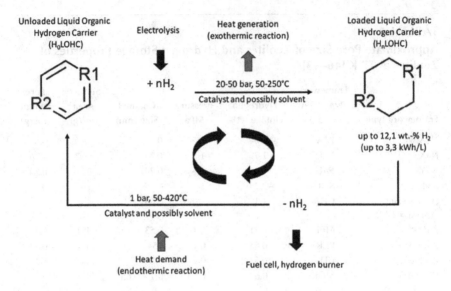

FIGURE 1.9   LOHC storage [55].

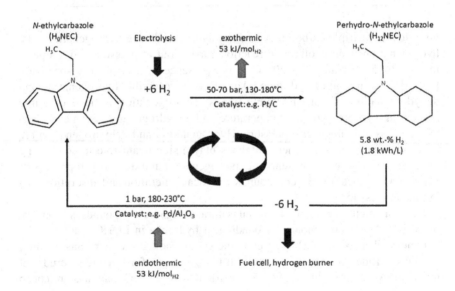

FIGURE 1.10   Storage process of N-ethyl carbazole [55].

Other LOHCs are dibenzyl toluene, 1,2-dihydro-1,2-azaborine, methanol, formic acid, etc. having similar properties with reasonably higher storage densities than other material-based storage and a promising option as a hydrogen carrier.

The major advantage of using LOHC is they are easily transportable via long distances. Moreover, they allow for long-term energy storage without boil-off or other hydrogen losses as well as uncomplicated transportation. LOHCs such as N-ethyl carbazole can be hydrogenated/dehydrogenated when the energy is needed.

1.4 HYDROGEN STORAGE REQUIREMENT FOR FUEL CELL VEHICLE

For the successful commercialization of fuel cell vehicles, safe and efficient storage of hydrogen is required for onboard applications. The technical and economic challenges of hydrogen storage system for onboard applications are major barriers to its acceptance. The first requirement of hydrogen fuel cell vehicles is to store 4–5 kg of hydrogen in lightweight tank in order to achieve a drive range of more than 500 km. The gravimetric energy density of hydrogen is extremely high but the volumetric storage density of the lightweight gas is low. For this, refueling of hydrogen plays an important role to increase the volumetric storage density. The inherently safer and efficient refueling can meet this demand of hydrogen storage system. This includes maintaining the temperature and pressure within the prescribed limit defined by SAE protocols.

The economic aspects such as cost of dispensed hydrogen, materials for lightweight tank and tank configuration are second important requirements for onboard applications. The cost of materials used in lightweight tank such as polymer, carbon fiber, and glass fiber is very high which increases the overall cost of the storage system. The composite materials generally contribute approximately 60% of the total cost of the tank [57].

Apart from this, rapid commercialization also needs regulations and policy frameworks for any technology. Many governmental and non-governmental authorities, mainly the US DOE, EU, SAE, ISO and ASME are actively involved in the development of codes, standards and policy frameworks. Countries like the USA, Japan and the EU have started offering various incentive programs for hydrogen fuel cell vehicles to address the various challenges in infrastructure development. The United States government has planned a hydrogen roadmap for legal policies and initiatives that are required to reduce carbon and NOx emissions by 16% and 36%, respectively [58]. The United States provides incentive benefits, credits, rebates and finance for fuel cell vehicles.

Japan is mainly dependent on importing fossil fuels to fulfill its energy demands. However, Japan has its ambitious goal on a mix of renewable energy, nuclear energy and fossil fuels and aims to reduce carbon emissions by 80% by 2050 [59]. For this, Japan is actively promoting hydrogen as fuel in various applications and is easing the legal framework to boost incentives and encourage the use of the fuel cell vehicles. Japan already has the highest incentives for fuel cell cars in the world and as a leader in hydrogen technology, the Japanese Ministry of Economy, Trade, and Industry (METI) published a strategic roadmap for hydrogen and fuel cells. These roadmaps are used to achieve a carbon-free society [60].

Similarly, the European Union is also committed to decarbonizing energy systems throughout Europe in order to align with the targets defined in the Paris agreement of 2016. The EU is planning to cut carbon emissions by 95% by 2050 [61]. To achieve these goals, the EU requires implementing hydrogen technologies on a wider scale. In the EU, the transport sector comprises one-third of the total carbon emissions. Therefore, decarbonizing the transport industry is a vital step to meet the standards of the Paris agreement [62]. In order to simplify the use of hydrogen and the development of this technology, a total of 25 member states of the EU signed the Hydrogen Initiative even before the EU hydrogen roadmap was initiated.

1.5 SUMMARY

The Chapter presents a detailed discussion on various hydrogen storage technologies. The discussion includes physical and materials-based storage systems. The simple operational features, low cost and less energy demand of compressed hydrogen make it a preferable option for onboard storage application. However, the storage density of compressed hydrogen storage is very less than the other physical methods. The liquid and cryo-compressed hydrogen storage has higher energy density but suffer from boil-off phenomenon. The high energy demand and cost of storage also restrict the liquid and cryo-compressed hydrogen. The materials-based storage methods discussed in this chapter have limitations of high-temperature requirement during charging and discharging of hydrogen. However, all methods have significantly higher storage density.

REFERENCES

1. Breeze, P., Chapter 8 - *Hydrogen energy storage, in power system energy storage technologies*, P. Breeze, Editor, Academic Press. pp. 69–77, 2018.
2. Abe, J.O., A.P.I. Popoola, E. Ajenifuja, O.M. Popoola, Hydrogen energy, economy and storage: Review and recommendation. *International Journal of Hydrogen Energy*, 44(29), 15072–15086, 2019.
3. Ren, J., Nicholas M. Langmi, Henrietta W. Mathe, Mkhulu Liao, Shijun, Current research trends and perspectives on materials-based hydrogen storage solutions: A critical review. *International Journal of Hydrogen Energy*, pp. 1–23, 2016.
4. Hydrogen Staorage Technologies Road Map US DOE, Editor. 2017: U.S. DRIVE (Driving research and innovation for vehicle efficiency and energy sustainability) https://www.energy.gov/sites/prod/files/2017/08/f36/hdtt_roadmap_July2017.pdf. Accessed on January 2019.
5. Leon, D.A., *Green energy and technoloygy (Hydrogen technology for mobile application)*. Springer-Verlag Berlin Heidelberg, 2008. (e-ISBN: 978-3-540- 69925-5).
6. Rosen, M.A., S. Koohi-Fayegh. The prospects for hydrogen as an energy carrier: An overview of hydrogen energy and hydrogen energy systems. *Energy, Ecology and Environment* 1, 10–29, 2016.
7. Aceves, S. M., G. Petitpas, F. Espinosa-Loza, M. J. Matthews, & E. Ledesma-Orozco, Safe, long range, inexpensive and rapidly refuelable hydrogen vehicles with cryogenic pressure vessels. *International Journal of Hydrogen Energy*, 38(5), 2480–2489, 2013.

8. Cardella, U., L. Decker, H. Klein. Economically viable large-scale hydrogen lique-faction. *IOP Conference Series: Materials Science and Engineering*, 171, 012013, 2017.

9. Klebanoff, L., Hydrogen storage in pressure vessels: Liquid, cryogenic, and com-pressed gas. In *Hydrogen storage technology materials and applications*, B. Bowman, Editor. Taylor & Francis Group, pp. 65–90, 2013.

10. Ahluwalia, R.K., J.K. Peng, T.Q. Hua, 5 - Cryo-compressed hydrogen storage. In *Compendium of hydrogen energy*, R.B. Gupta, A. Basile, T.N. Veziroğlu, Editors. Woodhead Publishing, pp. 119–145, 2016.

11. Barthelemy, H., M. Weber, F. Barbier. Hydrogen storage: Recent improvements and industrial perspectives. *International Journal of Hydrogen Energy*, 42(11), 7254–7262, 2017.

12. Stetson, N.T., S. McWhorter, C.C. Ahn, 1 - Introduction to hydrogen storage. In *Compendium of hydrogen energy*, R.B. Gupta, A. Basile, T.N. Veziroğlu, Editors. Woodhead Publishing, pp. 3–25, 2016.

13. Xianqiang Pei, M.E.K.F., Mechanical and tribological properties of PET/HDPE MFCs. *International Journal of Polymeric Materials*, 61(12), 963–977, 2012.

14. Liu, P.F.C., J.K. Hou, S.J. Xu, P. Zheng, Numerical simulation and optimal design for composite high-pressure hydrogen storage vessel: A review. *Renewable and Sustainable Energy Reviews*, 16(4), 1817–1827, 2012.

15. Ministry of New and Renewable Energy, GOI, New Delhi Hydrogen Energy and Fuel Cells in India – a way forward. 2016, Ministry of New and Renewable Energy, Government of India, New Delhi. http://mnre.gov.in/filemanager/UserFiles/Draft-Report-on-Hydrogen-Energy-and-Fuel-Cells-AWay-Forward.pdf. Accessed on July 2017.

16. Saba Niaz, T.M., Altaf Hussain Pandithn, Hydrogen storage: Materials, methods and perspectives. *Renewable and Sustainable Energy Reviews*, 50, 457–469, 2015.

17. Sakintuna B, F. Lamari-Darkrim, M. Hirscher, Metal hydride materials for solid hydrogen storage: A review. *International Journal of Hydrogen Energy*, 32, 1121–1140, 2007.

18. Graetz, J., J.J. Reilly. Kinetically stabilized hydrogen storage materials. *Scripta Materialia*, 56, 835–839, 2007.

19. Hauback, B.C. Structures of aluminium-based light weight hydrides. *Zeitschrift fuer Kristallographie*, 223, 636–648, 2008.

20. Graetz, J. 2009. New approaches to hydrogen storage. *Chemical Society Reviews*, 38, 73–82.

21. Schüth F, B. Bogdanovic, M. Felderhoff. Light metal hydrides and complex hydrides for hydrogen storage. *Chemical Communications* 2249–2258, 2004.

22. Orimo, S, Y. Nakamori, J.R. Eliseo, A. Züttel, C.M. Jensen. Complex hydrides for hydrogen storage. *Chemical Reviews*, 107, 4111–4132, 2007.

23. Bogdanovic, B., M. Schwickardi. Ti-doped alkali metal aluminium hydrides as poten-tial novel reversible hydrogen storage materials. *Journal of Alloys and Compounds*, 253–254, 1–9, 1997.

24. Jensen, C.M., K.J. Gross, Development of catalytically enhanced sodium aluminium hydride as a hydrogen-storage material. *Applied Physics A: Materials Science & Processing*, 72, 213–219, 2001.

25. Chen, P, Z. Xiong, J. Luo, J. Lin, K.L. Tan, Interaction of hydrogen with metal nitrides and imides. *Nature*, 420, 302–304, 2002.

26. Hino, S., T. Ichikawa, N. Ogita, M. Udagawa, H. Fujii. Quantitative estimation of NH3 partial pressure in H2 desorbed from the Li-N-H system by Raman spectroscopy. *Chemical Communications*, 3038–3040, 2005.

27. Nakamori, Y., S. Orimo, Borohydrides as hydrogen storage materials. In *Solid-state hydrogen storage: Materials and chemistry*, G. Walker, Editor, Woodhead Publishing, Cambridge, 2008.

28. Eberle, U., M. Felderhoff, F. Schüth. Chemical and physical solutions for hydrogen storage. *Angewandte Chemie, International Edition*, 48, 6608–6630, 2009.

29. Liu, Y., Pan, H., Chapter 13 - hydrogen storage materials. In *New and future developments in catalysis*, S.L. Suib, Editor, Elsevier, Amsterdam, pp. 377–405, 2013.

30. Yürüm, Y., A. Taralp, T.N. Veziroglu, Storage of hydrogen in nanostructured carbon materials. *International Journal of Hydrogen Energy*, 34, 3784–3798, 2009.

31. Huang, W.Z., X.B. Zhang, J.P. Tu, F.Z. Kong, J.X. Ma, F. Liu, *Materials Chemistry and Physics*, 78, 144–148, 2003.

32. Heine, T., L. Zhechkov, G. Seifert, *Physical Chemistry Chemical Physics*, 6, 980–984, 2004.

33. Pradhan, B.K., A. Harutyunyan, D. Stojkovic, P. Zhang, M.W. Cole, V. Crespi, Marine research society symposium proceedings. In *Making functional materials with nanotubes*, P. Bernier, P. Ajayan, Y. Iwasa, P. Nikolaev, Editors, vol. 706; p. Z10.3, 2002.

34. Zhou, L., J.S. Zhang, Y.P. Zhou, *Langmuir*, 17. 5503–5507, 2001.

35. Chambers, A., C. Parks, R.T.K. Baker, N.M. Rodriguez, *The Journal of Physical Chemistry B*, 102, 4253–4256, 1998.

36. Yin, Y.F., T. Mays, B. McEnaney. *Langmuir*, 16, 10521–10527, 2000.

37. Anson, A., M.A. Callejas, A.M. Benito, W.K. Maser, M.T. Izquierdo, B. Rubio. *Carbon*, 42, 1243–1248, 2004.

38. Li, G., L. Xia, J. Dong, Y. Chen, Y. Li, *10* - Metal-organic frameworks. In *Solid-phase extraction*, C.F. Poole, Editor, Elsevier, pp. 285–309, 2020.

39. Sagara, Tatsuhiko, James Klassen, Eric Ganz, Computational study of hydrogen binding by metal-organic framework-5. *The Journal of Chemical Physics*, 121, 12543, 2004.

40. Urukawa, H., M.A. Miller, O.M. Yaghi, Independent verification of the saturation hydrogen uptake in MOF-177 and establishment of a benchmark for hydrogen adsorption in metal–organic frameworks. *Journal of Materials Chemistry*, 17, 3197–3204, 2007.

41. Wong-Foy, A.G., A.J. Matzger, O.M. Yaghi, Exceptional H2 saturation uptake in microporous metal organic frameworks. *Journal of the American Chemical Society*, 128, 3494–3495, 2004.

42. Hans, S., W. Goddard, Lithium doped metal organic frameworks for reversible H2 storage at ambient temperature. *Journal of the American Chemical Society*, 129, 8422–8423, 2007.

43. El Kassaoui, M., M. Lakhal, A. Benyoussef, A. El Kenz, M. Loulidi, Enhancement of hydrogen storage properties of metal-organic framework-5 by substitution (Zn, Cd and Mg) and decoration (Li, Be and Na). *International Journal of Hydrogen Energy*, 46(52), 26426–26436, 2021.

44. Dong, J., X. Wang, H. Xu, Q. Zhao, J. Li, Hydrogen storage in several microporous zeolites. *International Journal of Hydrogen Energy*, 32(18), 4998–5004, 2007.

45. Adriano Zecchina, Silivia Bordiga, Vitillo Jenny G. Liquid hydrogen in protonic chabazite. *Journal of the American Chemical Society*, 127, 6361–6366, 2005.

46. Nijkamp, M.G., J.E.M.J. Raaymakers, A.J. van Dillen, K.P. Jong, Hydrogen storage using physisorption-materials demands. *Applied Physics A: Materials Science & Processing*, 72, 619–923, 2001.

47. Langmi, H.W., A. Walton, M.M. Al-Mamouri, Hydrogen adsorption in zeolites A, X, Y and RHO. *Journal of Alloys and Compounds*, 356–357, 710–715, 2003.

48. Sheppard, D.A., C.F. Maitland, C.E. Buckley, Preliminary results of hydrogen adsorption and SAXS modeling of mesoporous silica: MCM-41. *Journal of Alloys and Compounds*, 404–406, 405–408, 2005.

49. van den Berg, A.W.C., S.T. Bromley, J.C. Wojdel, J.C. Jansen. Adsorption isotherms of H2 in microporous materials with the SOD structure: A grand canonical Monte Carlo study. *Microporous and Mesoporous Materials*, 87, 235–242, 2006.

50. Baerlocher, Ch, W.M. Meier, D.H. Olson, *Atlas of zeolite framework types*. Elsevier, Amsterdam, pp. 190–191, 196–197, 140–141, 2001.

51. Makarova, M.A., V.L. Zholobenko, K.M. Alghefaili, N.E. Thompson, J. Dewing, J. Dwyer. Bronsted acid sites in zeolites. FTIR study of molecular hydrogen as a probe for acidity testing. *Journal of the Chemical Society, Faraday Transactions*, 90, 1047–1054, 1994.

52. Hartmann, M., C. Bischof, Z. Luan, Preparation and characterization of ruthenium clusters on mesoporous supports. *Microporous and Mesoporous Materials*, 44–45, 385–394, 2001.

53. Zheng, J., H. Zhou, C. G. Wang, E. Ye, J. W. Xu, X. J. Loh, Z. Li, Current research progress and perspectives on liquid hydrogen rich molecules in sustainable hydrogen storage. *Energy Storage Materials*, 35, 695–722, 2021.

54. Abdin, Z., C. Tang, Y. Liu, K. Catchpole, Large-scale stationary hydrogen storage via liquid organic hydrogen carriers. *iScience*, 24(9), 2021.

55. Niermann, M., A. Beckendorff, M. Kaltschmitt, K. Bonhoff, Liquid Organic Hydrogen Carrier (LOHC) – Assessment based on chemical and economic properties. *International Journal of Hydrogen Energy*, 44(13), 6631–6654, 2019.

56. Ahluwalia, R.K., T.Q. Hua, J.K. Peng, On-board and Off-board performance of hydrogen storage options for light-duty vehicles. *International Journal of Hydrogen Energy*, 37(3), 2891–2910, 2012.

57. Asif, U., K. Schmidt, Fuel Cell Electric Vehicles (FCEV): Policy advances to enhance commercial success. *Sustainability*, 13, 5149, 2021.

58. Iida, S., K. Ko Sakata, Hydrogen technologies and developments in Japan. *Clean Energy*, 3, 105–113, 2019.

59. Greenwood, M., Japan sees big future in hydrogen cars. 8 April 2019. Available online: https://new.engineering.com/story/japan-sees-big-future-in-hydrogen-cars. Accessed on May 2021.

60. Jong, W., G. Honselaar, O. Cebolla, Institute for energy and transport (Joint Research Centre). CEN—CENELEC sector forum energy management: Working group hydrogen: Final report. 21 December 2015. Available online: https://op.europa.eu/en/publication-detail/-/publication/99f62cea-a877-11e5-b528-01aa75ed71a1. Accessed on May 2021.

61. Fuel Cells and Hydrogen 2 Joint Undertaking. Hydrogen roadmap Europe: A sustainable pathway for the European energy transition. 15 February 2019. Available online: https://op.europa.eu/en/publication-detail/-/publication/0817d60d-332f-11e9-8d04-01aa75ed71a1/language-en. Accessed on May 2021.

2 Compressed Hydrogen Storage

2.1 INTRODUCTION

Hydrogen carries larger amount of energy than conventional sources like gasoline and diesel. The utilization of hydrogen energy requires a distinct infrastructure. The various elements of the infrastructure such as production, delivery, storage, fuel cell, vehicles, codes and standards, safety, technology validation and education and their development comprise "hydrogen economy" (Figure 2.1). Among all, "storage" of hydrogen plays a pivotal role in triggering the acceptance of hydrogen as the fuel of the future.

For the success of hydrogen economy, safe and efficient means of storing hydrogen are required for on-board, portable and stationary applications. Hydrogen storage, across production, delivery and end-use applications is an undaunted task [1]. Indeed, for a successful application of hydrogen as an energy carrier, it should be stored safely and efficiently as conventional fuels [2]. Hydrogen storage is a key process that enables the technology for the successful penetration of hydrogen fuel cell vehicles in the market dominated by gasoline vehicles [3].

Size, weight and density are some deciding factors for any storage system. However, sometimes, the characteristics of fuel also play a significant role in defining the suitability of the storage system. Hydrogen possesses unusual physical and thermodynamic properties that make it difficult to call any storage method as efficient and full-proof for particular applications. Under normal temperature and pressure, the density of the H_2 gas is very low, near about 0.08238 kg/m^3; e.g. for storing 5 kg of hydrogen, which implies a volume of around 60 m^3 and energy content of 600 MJ (166.65 kWh). For the same weight and energy content, the required gasoline volume is 0.019 m^3 [4].

In view of these numbers, it is clear that for efficient storage, hydrogen density should be increased by reducing the volume taken by the gas under normal temperature and pressure conditions. Therefore, the "normal state" of hydrogen has to be changed in order to store it efficiently. This can be accomplished by increasing the pressure, decreasing the temperature below the critical temperature or reducing the repulsion interaction between hydrogen molecules by binding them with another material [5].

For this purpose, hydrogen is stored as a high-pressure gas called compressed hydrogen (CH_2) or low-temperature liquid called liquid hydrogen (LH_2). In the first approach, hydrogen is stored as a compressed gas at room temperature

DOI: 10.1201/9781003244318-2

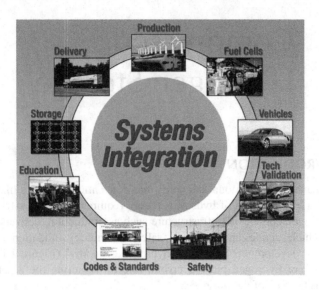

FIGURE 2.1   Hydrogen Economy [1].

conditions, in a metal or composite cylinder with a maximum pressure of up to 70 MPa [5].

The liquid hydrogen storage is realized by liquifying it at the temperature of –253°C (21 K) and ambient pressure conditions. Due to the low critical temperature of hydrogen (–240°C or 33 K), it has to be liquified lower than the critical temperature for attaining the liquid phase. The storage density achieved by liquification is higher than the compression. But, liquification needs 30% energy of the lower heating value of hydrogen [6]. Apart from this, the boil-off phenomenon further demands an extra energy input after a certain period of storage. Hence, energy consumption and leakage along with safety issues restricted this solution to special applications only.

Currently, "cryo-compressed" technique is the latest approach for storing the hydrogen at a pressure above normal state and temperature similar to or less than liquid nitrogen. It is a combination of two approaches where hydrogen is compressed to higher pressure up to 35 MPa and cooled to temperature –233°C. When the pressure of liquified hydrogen is increased the storage density increases to 90 g/L [6]. BMW, an automotive OEM, has demonstrated this technique in its fuel cell vehicle BMW-7. Table 2.1 presents the comparison of storage densities of compressed, liquid and cryo-compressed storage systems.

Compressed hydrogen storage has the lowest density compared to others but it is a commercially acceptable and mature technology. The compression of hydrogen up to 70 MPa consumes 1–3 kWh/kg which is much less than the energy required for liquification of hydrogen. It consumes more than 10 kWh/kg energy for liquification with approximately 30% loss of chemical energy of hydrogen, which makes liquified hydrogen an expensive matter. Therefore, compressed

TABLE 2.1

Physical Hydrogen Storage Methods [5,7]

Hydrogen Storage Methods	Storage Density (g/L)	Operating Pressure and Temperature
Compressed	40	70 MPa, 15°C
Liquid	71	0.1 MPa, −253°C
Cryo-compressed	90	35 MPa, −233°C

hydrogen has become a more promising option in terms of low energy consumption and less cost of the storage system for onboard applications.

2.2 COMPRESSED HYDROGEN TANKS

Storage of compressed hydrogen at high operating pressure requires a highly stable system which can withstand cyclic loading conditions during its life cycle. For this purpose, high-pressure hydrogen is stored in thick-walled tanks made of high-strength material to ensure durability. According to the American Society of Mechanical Engineers (ASME) and International Organization for Standardization (ISO) hydrogen storage tanks are classified into four categories (Figure 2.2).

Type I tanks are metallic cylinders with operating pressure range limited to 20 MPa; generally preferred for stationary applications.

Type II tanks are thick-wall metallic cylinders wrapped with fiber resin composites, exclusively for the cylindrical part of the tank, and have a similar operating pressure range as Type I tank; generally preferred for stationary application.

Type III tanks are made of metallic liner and fiber resin composite over-wrapped along the surface of the cylinder. They can be used for higher pressure range up to 45 MPa which is suitable for onboard application of the tank.

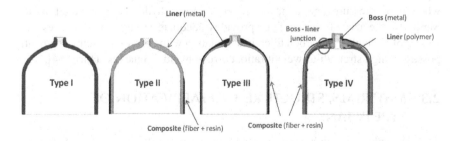

FIGURE 2.2 Schematic of Type I, Type II, Type III and Type IV tanks [8].

TABLE 2.2

Comparison of Performance and Cost of the Compressed Hydrogen Storage Systems [11]

Compressed hydrogen storage tank	Gravimetric Density (kWh/kg)	Volumetric Density (kWh/L)	Cost ($/kWh Project for 50,00,000 Units/Year)
Type IV (70 MPa)	1.40	0.81	14.80
Type III (70 MPa)	0.83	0.80	30.00

> **Type IV** tanks are the polymer-lined tank overwrapped with fiber resin composite. Due to the high strength, composite wrapping the operating pressure range increases to more than 100 MPa. Presently, a Type IV tank is preferred for onboard application.

The most advanced lightweight storage system for onboard application is actually a composite tank with non-load carrying metallic (Type III) or polymer (Type IV) liner composite tanks. Type III and IV tanks have attracted much interests from the scientific community and also become the first choice for fuel cell vehicle manufacturers. Their structure is based on two fundamental components: the liner, essentially a barrier for hydrogen permeation, and the composite structure that ensures the mechanical integrity of the tank. However, the metallic embrittlement of the liner at high pressure caused the restriction of the use of Type III tank [9]. The Type IV tank generally has a combination of plastic liner made of high-density polyethylene (HDPE), polyamide and carbon fiber (CF) with epoxy resins which reduce the weight of the tank with a higher load-carrying capacity [10]. This leads to the additional advantage for Type IV tank to be used in fuel cell vehicle application. Apart from this, the performance and cost of compressed hydrogen in Type IV tank carries a serious advantage over Type III tank, as per tabulated data in Table 2.2. Hence, the Type IV tank at pressures of 35 MPa and 70 MPa are relatively more promising options due to their high strength/stiffness-to-weight ratio, and excellent resistance to fatigue and corrosion [12].

The HDPE has high tensile/compressive strength and stiffness even at high pressure and temperature [13]. Therefore, it acts as good permeation resistance when high pressure and temperature compressed hydrogen are stored in it. Similarly, the use of carbon fiber provides adequate strength and reduces the weight of the tank. The carbon fiber, in association with epoxy resin reinforcement, possesses high strength-to-weight ratio, corrosion and fatigue resistance [14].

2.3 MATERIALS, STRUCTURE AND FABRICATION OF TYPE IV TANK

Type IV tanks are mainly composed of three layers of different materials to provide sufficient strength against severe internal and external loading conditions. The inner layers are made of HDPE liner which acts as a hydrogen barrier and

TABLE 2.3

Mechanical and Thermal Properties of HDPE Liner [13,15]

Parameter	Mechanical Properties	Thermal Properties
Density	0.93–0.965 g/m³	–
Modulus of elasticity	1035 N/mm²	–
Poisson ratio	0.40–0.45	–
Specific heat	–	2.40–2.45 kJ/kg-°K
Thermal conductivity	–	140–232 kJ/kg
Coefficient of linear thermal expansion	–	12×10^{-5} (μm/m · K)

provides the shell for overwrapping the outer layer. The mechanical and thermal properties of HDPE make it the first choice for Type IV manufacture (as in Table 2.3). HDPE also provides thermal stability and toughness to the tank structure against a large number of charging and discharging cycles. It also resists the impact, fatigue and chemical loads, even at cryogenic temperatures. Currently, HDPE is mostly used as a liner by many manufacturers like Hexagon Composite, Quantum Technologies and Lincoln, by virtue of its high thermal stability up to 120°C and excellent chemical resistance.

The liner of a composite pressure vessel is commonly fabricated using injection, blow, compressive, and rotational molding processes. All these processes are being utilized to produce hard and high-strength plastic products for commercial use. In injection molding, melted plastic is forced in the mold cavity which is, in general, preferred for small products in mass production units. Blow molding is similar to injection except the melted plastic pours out of the barrel vertically in a molten tube and cooled to form a hollow part. The process is used for making tubes, small containers and bottles. In compressive molding, hard plastic is pressed against the heated molds than air-cooled.

When plastic is rotated in hollow molds and cooled by water sprayed to cause the plastic to harden and obtained as a hollow part called rotational molding. Rotation molding involves a bi-axial rotation of metallic mold in a heated oven for producing stress-free parts.

The powdered plastic is heated at high temperature in a mold which rotates on a perpendicular axis. The heating time, temperature and rotational speed are major parameters that affect the mechanical strength of the liner [16]. The heated materials are then cooled down by water spray at a certain cooling rate to achieve the required solidification followed by easy removal (Figure 2.3). Air cooling is preferred over water cooling, particularly to avoid shrinking [17,18]. This method is less complicated with a wide range of surface finishes and the product obtained is free from residual stress.

Rotational molding is less expensive, low waste and preferred for low- to high-volume production. Compared to other molding processes such as injection, extrusion molding, rotational molding is more ideal for fabricating hollow shapes in a variety of sizes, which results in better part-wall-thickness distribution [18]. However,

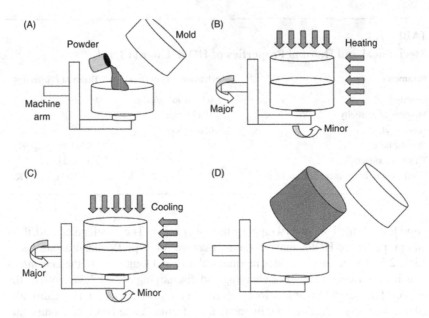

FIGURE 2.3 Rotational molding steps (A) Loading (B) Heating (C) Cooling (D) Unloading [16].

rotational molding requires materials having good flow-ability, i.e. very low viscosity at zero-shear-rate; a lot of materials with excellent barrier properties cannot be used due to relatively high viscosity, i.e. polyethylene terephthalate (PET) [13].

The outer shell of Type IV tank is made of fiber-reinforced polymer with epoxy resin. Table 2.4 has listed the common composite materials used in Type IV tank by various manufacturers and their mechanical properties.

TABLE 2.4
Mechanical Properties of Materials for Outer Shell [19,20]

Fiber type	Tensile Strength (MPa)	Tensile Modulus (GPa)	Density g/cm²	Specific heat Capacity J/g·°C	Thermal Conductivity: J/cm·s·°C.	Composite Strength MPa
T300 Carbon	3,530	230	1.76	0.777	0.105	
T700S Carbon	4,900	230	1.80	0.752	0.096	2,830
T700G Carbon	4,900	240	1.80	0.752	0.096	
T800	5,880	294	1.80	0.752	0.113	
Flax	8.25	23	1.45	–	–	345
Basalt	50	80	1.80	–	–	1,600

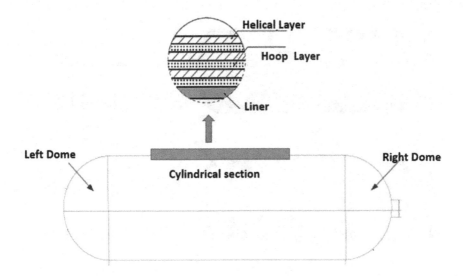

FIGURE 2.4 Helical and hoop winding on the cylindrical shell of composite tank.

Among these materials, carbon fiber with epoxy resin is generally preferred for Type IV tanks due to its strength, flexibility and higher translation efficiency. However, as far as cost is concerned, carbon fiber composite tanks are found a bit expensive. Carbon fiber contributes almost 62% of total weight of the tank and has a higher per kg cost. A composite layer of Type IV tank is a combination of a carbon fiber wrapped in a hoop and helical direction to provide strength to the structure (Figure 2.4). The placement or wrapping of fiber is commonly performed by the filament winding technique.

Filament winding is a very popular method to produce a symmetric composite product such as tanks, tubes and cylinders, and domes. In this method, carbon fibers are deposited in a specifically oriented pattern that matches the direction of stresses and load in the structure [21]. Fibers are placed on polymer shells in the formed layers called ply at a defined angle, known as ply angle in helical and hoop directions as shown in Figure 2.5.

The placement of fiber in helical direction can also be divided as low-, and high-angle helical winding to support the junction of cylindrical and dome sections. Low-angle helical winding is generally used to strengthen the dome regions in the axial direction. However, the high-angle helical winding is required to strengthen the boundary regions (Figure 2.6).

The fibers are wound in a hoop and helical direction forming a ply or lamina precisely using mandrel (Figure 2.7). In the typical process, the fiber processed from fiber strands to resin basin and impregnated fiber moves forward to the rotating mandrel of the filament winding machine. The impregnated fiber wound around the mandrel in a controlled manner with specific tension in fiber, orientation and speed (pulling the fiber through the basin of resin) [23].

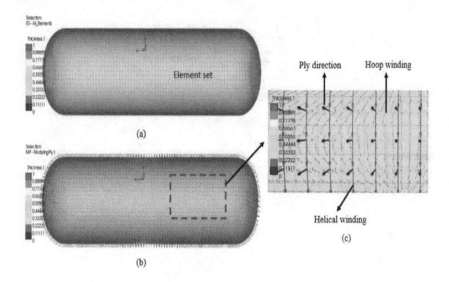

FIGURE 2.5   Deposition of carbon fiber layer and ply in the composite shell of Type IV tank.

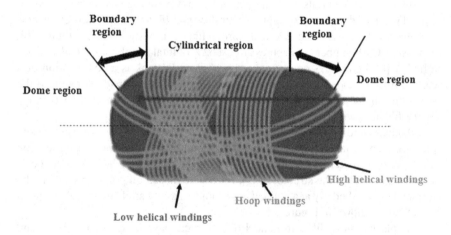

FIGURE 2.6   Winding pattern on cylindrical and dome section of the tank [22].

The filament winding provides deposition of fibers along the direction of stress and loads to ensure the maximum utilization of high strength of fibers and eventually increases structural stability [21]. The major advantage with the filament winding approach is the production of high-strength products due to continual control over fiber angle. The method is prone to automation for high production volume. The low external surface finish and complex shapes are some limitations of the filament winding process.

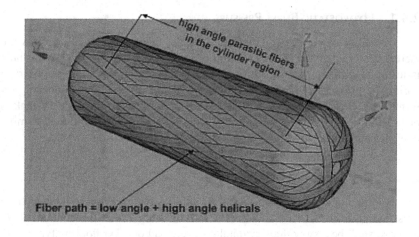

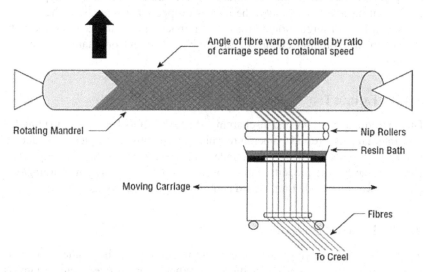

FIGURE 2.7 Deposition of fiber using filament winding process.

Source: https://netcomposites.com/guide/manufacturing/filament-winding/

2.4 TESTING OF TYPE IV TANK

The compressed hydrogen storage tank has to undergo various testing procedures to ensure its safety for application in fuel cell vehicles. According to ISO/TS 15869, the safety of the tank is one of the critical issues that has to be addressed for its public acceptance. The leakage, overpressure (burst), vehicle fire, liner issues and accidental failure are some of the major reasons for the failure of a storage tank. For identification of possible failure, various tests such as burst, leak detection, cycling, drop test, gunfire test, extreme temperature test, etc. are recommended by SAE J2760 and ISO 15869 to ensure the safety of the tank [24].

2.4.1 Hydrostatic Burst Pressure

The durability of composite tank tested by the failure of the tank as pressure increases more than the prescribed limit of safe storage by SAE J2601. This tank is tested at different pressurization rates for fixed time and strain is observed during the pressurization. The test for performance durability and onboard performance follows the SAE J2579 [25]. During the testing, the Type IV tank is pressurized to 80–150 times of nominal working pressure for a different number of cycles and times.

2.4.2 Drop (Impact Test)

The storage container is drop tested at ambient temperature without internal pressurization or attached valves. The surface onto which the containers are dropped shall be a smooth, horizontal concrete pad or other flooring types with equivalent hardness. Generally, the tank is dropped at 1.8 m above the ground surface at four different orientations. The impact on the tank is measured with respect to the potential energy of the tank at different orientations from the ground.

2.4.3 Fatigue Test

Due to frequent charging and discharging, the tank suffers from cyclic loading for high internal pressure and temperature which lead to fatigue. This test is conducted to identify fatigue behavior of tank under hydrogen environment. The fatigue test identifies the fatigue property, failure behavior, and safe hydrogen charging/discharging working mode for onboard composite hydrogen storage vessels.

2.4.4 Extreme Temperature

Tank is exposed to high-temperature environment or open fire to check the thermal response from materials of the tank. When a pressurized vehicle tank is exposed to excessive heat arising from a vehicle fire, the tank materials begin to degrade and the pressure increases. The SAE Fuel Cell Safety Committee developed a combined localized/engulfing fire test that was eventually adopted by the UN GTR 13 regulation, "Global Technical Regulation on Hydrogen and Fuel Cell Vehicles" [26]. The development of the localized/engulfing fire test procedure is described in UN GTR 13 which reported that at the temperature of 300°C where the localized fire condition could start and materials begin to degrade rapidly.

2.4.5 Gunfire Penetration Test

In this test, the tank is positioned in such a way that the impact point is in the cylinder side wall at a 45° angle with respect to the longitudinal axis of the cylinder. The distance from the firing location to the cylinder may not exceed 45.7 m (150 ft) of fire penetration. The tank is tested for impact strength by penetration of a bullet, fired to the tank.

Apart from these essential tests, some tests such as environmental, permeation, hydrogen cycling, and tensile test are also preferred by some manufacturers.

2.5 CODES AND STANDARDS FOR TYPE IV TANK

The safety of hydrogen stored at high pressure is a major concern for the storage tank manufacturers and suppliers. For this, various organizations such as International Organization for Standardization (ISO), American National Standards Institute (ANSI), Compressed Gas Association (CGA), National Fire Protection Association (NFPA), American Society of Mechanical Engineers (ASME), European Committee for Standardization (CEN), Society of Automotive Engineers (SAE) are involved in the development of guidelines and regulation for compressed hydrogen storage tanks.

ASME section X defines the standard for materials required for composite Type IV tank manufacturing such as fiber, resins, and composite. Section X also covers the design requirement for composite pressure vessels such as design pressure, loading, restrictions, allowance and limitations.

ISO 15869:2009 specifies the requirements for lightweight refillable fuel tanks intended for the onboard storage of high-pressure compressed gaseous hydrogen [27–29]. It is applicable for metallic cylinders mainly made of steel, stainless steel, aluminium, polymer and composite materials. ISO 15869:2009 applies to the following types of fuel tank designs:

a) Type I: metal fuel tanks;
b) Type II: hoop-wrapped composite fuel tanks with a metal liner;
c) Type III: fully wrapped composite fuel tanks with a metal liner;
d) Type IV: fully wrapped composite fuel tanks with no metal liner.

TABLE 2.5
General Design and Safety Standards for Hydrogen Storage [29]

S.No.	Standard	Name of Standard
1	CGA H-5-2014	Installation standard for bulk hydrogen supply systems
2	CGA P-12-2017	Safe handling of cryogenic liquids
3	CGA PS-17-2004	CGA position statement on underground installation of liquid hydrogen storage tanks
4	CGA PS-20	CGA Position statement on the direct burial of gaseous hydrogen storage tanks
5	NFPA-2	Gaseous storage technology code
6	NFPA-55	Compressed gas code

CGA provides eight standards for the design and safety of storage of hydrogen as listed in Table 2.5. Among them, CGA PS-20 is specifically defined for compressed hydrogen storage. NFPA covers the safety issues against fire in on-board and off-board storage of hydrogen in its article NFPA 2 and 55 [29].

CEN in their code for gaseous cylinders CEN/TC 23 (Transportable gas cylinders) covers the standardization of transportable gas cylinders, their fittings, and requirements relating to their design testing and operation. Apart from these codes and standards, ISO and SAE have developed a global technical report (GTR-13) for the requirements of hydrogen storage systems for fuel cell vehicles [26].

2.6 SUMMARY

The storage of hydrogen in compressed form is a feasible solution due to less weight as compared to other metallic tanks. The polymer and carbon fiber provide hydrogen barrier and enough strength respectively, to Type IV tank to sustain at high pressure of compressed hydrogen. The ease of fabrication of Type IV tanks made them a preferred choice for fuel cell vehicle manufacturers. However, the high cost of materials is still a great challenge to be dealt with. The required ongoing development of standards for onboard and stationary applications will accelerate the acceptance of fuel cell vehicles.

REFERENCES

1. Hydrogen Production Cost Estimate Using Biomass Gasification. U.S. National renewable energy laboratory. October 2011. (https://www.hydrogen.energy.gov/pdfs/51726.pdf) Accessed on August 2020.
2. Breeze, P., Chapter 8 - hydrogen energy storage. In *Power system energy storage technologies*, P. Breeze, Editor. Academic Press, Elseveir USA, pp. 69–77, 2018.
3. Abe, John O., A.P.I. Popoola, Emmanueal Ajenifuja, O.M. Popoola, Hydrogen energy, economy and storage: Review and recommendation. *International Journal of Hydrogen Energy*, 44(29), 15072–15086, 2019.
4. Broom, D., *Hydrogen storage materials: The characterization and their storage properties*. Springer-Verlag London Limited, London, 2011.
5. Leon, D.A., *Green energy and technoloygy (Hydrogen Technology for mobile application)*. Springer-Verlag Berlin, Heidelberg, 2008.
6. Klebanoff, L., Hydrogen storage in pressure vessels: Liquid, cryogenic, and compressed gas. In *Hydrogen storage technology materials and applications*, B. Bowman Editor. Taylor & Francis Group, pp. 65–90, 2013.
7. Hydrogen Staorage Technologies Road Map U. DOE, Editor. U.S. DRIVE (Driving research and innovation for vehicle efficiency and energy sustainability). 2017. (https://www.energy.gov/sites/prod/files/2017/11/f46/HPTT%20Roadmap%20FY17%20Final_Nov%202017.pdf) Accessed on December 2019.
8. Barthélémy, H., M. Weber, F. Barbier, Hydrogen storage: Recent improvements and industrial perspectives. (https://h2tools.org/sites/default/files/2019-09/293.pdf) Accessed on January 2021.
9. Ross, D.K., Hydrogen storage: The major technological barrier to the development of hydrogen fuel cell cars. *Vacuum*, 80(10), 1084–1089, 2006.

10. Pei, Xianqiang, M. Evstatie, K. Friedrich, Mechanical and tribological properties of PET/HDPE MFCs. *International Journal of Polymeric Materials*, 61(12), 963–977, 2012.

11. Zhang, M., H. Lv, H. Kang, W. Zhou, C. Zhang, A literature review of failure prediction and analysis methods for composite high-pressure hydrogen storage tanks. *International Journal of Hydrogen Energy*, 44(47), 25777–25799, 2019.

12. Pareek, K., Q. Zhang, R. Rohan, Z. Yunfeng, H. Cheng, Hydrogen physisorption in ionic solid compounds with exposed metal cations at room temperature. *The Royal Society of Chemistry*, 04, 33905–33910, 2014.

13. Hamid, F., A. Suffiyana, K.K. Halim, Mechanical and thermal properties of polyamide 6/HDPE-g- MAH/high density polyethylene. *Procedia Engineering*, 68, 418–424, 2013.

14. Park, J.S., C.S. Hong, C.G. Kim, C.U. Kim, Analysis of filament wound composite structures considering the change of winding angles through the thickness direction. *Composite Structures*, 55(1), 63–71, 2002.

15. Hamid, F., A. Suffiyana, K.K. Halim, Mechanical and thermal properties of polyamide 6/HDPE-g- MAH/high density polyethylene. *Procedia Engineering*, 68, 418–424, 2013.

16. Nugent, Paul, *Applied plastics engineering handbook (Second Edition) processing, materials, and applications plastics design library*, Elsevier Science, pp. 321–343, 2017.

17. Neto, E.B., M. Chludzinski, P.B. Roese, J.S.O. Fonseca, S.C. Amico, C.A. Ferreira, Experimental and numerical analysis of a LLDPE/HDPE liner for a composite pressure vessel. *Polymer Testing*, 30(6), 693–700, 2011.

18. Barboza Neto, E.S., L.A.F. Coelho, M.M.D.C. Forte, S.C. Amico, C.A. Ferreira, Processing of a LLDPE/HDPE pressure vessel liner by rotomolding. *Materials Research*, 1, 236–241, 2014.

19. Bouvier, Mathilde, Vincent Guiheneuf, Alan Jean-Marie, Modeling and simulation of a composite high-pressure vessel made of sustainable and renewable alternative fibers. *International Journal of Hydrogen Energy*, 44(23), 11970–11978, 2019.

20. Yamashita, A., M. Kondo, S. Goto, N. Ogami, Development of high pressure hydrogen storage system for the Toyota "Mirai". *SAE Technical Paper 2015- 01-1169*; 2015.

21. Behera, S., S.K. Sahoo, L. Srivastava, A.S. Gopal, Structural integrity assessment of filament wound composite pressure vessel using through transmission technique. *Procedia Structural Integrity*, 14, 112–118, 2019.

22. Yamashita, A., M. Kondo, S. Goto, N. Ogami, Development of high pressure hydrogen storage system for the toyota "Mirai". *SAE Technical Paper 2015- 01-1169*; 2015.

23. Teng, T.L., C.M. Yu, Y.Y. Wu, Optimal design of filament-wound composite pressure vessels. *Mechanics of Composite Materials*, 41, 333–340, 2005.

24. Odegard, B.C., G.J. Thomas, Testing of high pressure hydrogen composite tanks. 2001 Sandia National Laboratories: DOE Hydrogen Program Review. (https://www1.eere.energy.gov/hydrogenandfuelcells/pdfs/30535bd.pdf) Accessed on January 2020.

25. Burgess, R., C. Blake, C.E. Tracy, Test protocol document, hydrogen safety sensor testing. 2008, NREL. (https://www.nrel.gov/docs/fy08osti/42666.pdf) Accessed on September 2019.

26. Buttner, W., C. Rivkin, R. Burgess, K. Hartmann, I. Bloomfield, M. Bubar, M. Post, L. Brett, E. Weidner Ronnefeld, P. Moretto, Hydrogen monitoring requirements in the global technical regulation on hydrogen and fuel cell vehicles. *ICHS, International Journal of Hydrogen Energy*, 42(11), 7664–7671, 2017, JRC106642, ISSN 0360-3199.

27. Joe Wong, P.E., DOE tank safety workshop hydrogen tank safety testing. 2010, POWERTECH –Hydrogen & CNG Services. (https://www.energy.gov/sites/prod/files/2014/03/f10/hydrogentank_testing_ostw.pdf) Accessed on December 2019.
28. Joe Wong, P.E., DOE Hydrogen tank testing program 2010, Power Tech. (https://www.powertechlabs.com/services-all/hydrogen-fuel-cell-vehicle-testing) Accessed on March 2020.
29. Yang, Yanmei, Haigang Xu, Qiaoling Lu, Wei Bao, Ling Lin, Bin Ai, Bangqiang Zhang, Development of standards for hydrogen storage and transportation. *EDP Sciences*, 194, 2020. https://doi.org/10.1051/e3sconf/202019402018

3 Compressed Hydrogen Refueling Stations

3.1 HYDROGEN REFUELING STATIONS

Hydrogen has been identified as a potential energy carrier. The extraction or utilization of hydrogen energy depends on the development of hydrogen infrastructure mainly, transportation, distribution and storage. For significant deployment of fuel cell vehicles, need of comprehensive refueling infrastructure would be required. As the market of fuel cell vehicles grows a sufficient number of refueling stations need to be developed. Therefore, the lack of adequate hydrogen delivery infrastructure is a major barrier to the large-scale deployment of hydrogen fuel cell vehicles. However, the presence of uncertainty between investment in hydrogen infrastructure and penetration of fuel cell vehicles is creating the chicken and egg dilemma.

The demand and supply of hydrogen can only be supported with the presence of sufficient infrastructure. The initial investment required for a hydrogen refueling station is very high and not profitable if it is underutilized [1]. Therefore, a public-private partnership is supported by many governments to initiate the deployment of hydrogen refueling stations and fuel cell vehicles. Many countries such as the United State of America, Japan, Germany and the United Kingdom have started this initiative to accelerate the development of hydrogen refueling station network [2].

Currently, the infrastructure for hydrogen fuel cell vehicles has been under development in various countries. Most of them are exclusively for the compressed hydrogen filling. Compressed hydrogen for fuel cell vehicles is a commercially acceptable technology by fuel cell vehicle manufacturers. Almost all OEMs agree that pressurized hydrogen at 70 MPa for on-board applications is an appropriate option to obtaining the drive range up to 500 km in a single fill [3]. The compressed hydrogen filling station operates on the concept of providing gas at the required pressure. The high-pressure hydrogen up to 100 MPa or more is stored at a refueling station reservoir for a nominal working pressure of 70 MPa at a fuel cell vehicle tank. The refueling station mainly consists of a storage and dispenser unit. However various additional mechanical systems are used to store and maintain the pressure of the hydrogen for a longer period of time. Generally, the storage tank of the fuel cell vehicle is connected to the refueling station reservoir and filled till the vehicle storage reached the target pressure.

Common configurations of filling of compressed hydrogen are shown in Figure 3.1. Where, in Figure 3.1(a), compressed hydrogen is supplied to vehicle storage from the

DOI: 10.1201/9781003244318-3

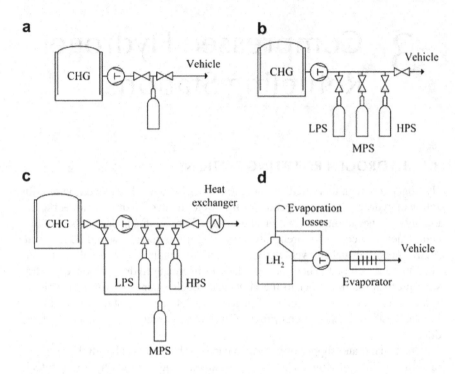

FIGURE 3.1   Filling station types: (a) single-stack system, (b) multiple-stack system, (c) multiple-stack system with booster and (d) cryogenic compressor [4].

reservoir at the refueling station having pressure more than 1.25 times of nominal operating pressure by simply connecting both [4].

Another arrangement of filling of vehicle storage is a multiple-stack storage system at refueling station as shown in Figure 3.1(b). In this storage, the vehicle tank is connected to station storage in successions of low, medium and high to reduce the pressure differential. This also reduces the temperature rise due to the Joule–Thomson effect during the expansion [4]. The pressure difference in the medium pressure storage tank can be overcome by using a booster compressor to increase the pressure in high-pressure storage as shown in Figure 3.1(c) [5]. The high-pressure compressed hydrogen increases the vehicle tank temperature which requires cooling.

For this compressed gas from the booster compressor is cooled in the refrigeration unit to maintain the gas temperature below 85°C inside the vehicle tank as per SAE J2601 [6]. This type of filling also known as booster filing and configuration is shown in Figure 3.2 [7]. The disadvantage of using the arrangement "c" is that the booster compressor plays an additional role in increasing the temperature of the gas inside the vehicle tank. Therefore, a cooling unit or heat exchanger with a chiller has to be applied to cool the compressed gas before dispensing it to the vehicle tank called the "precooling unit" [8].

The Society of Automotive Engineers (SAE) has categorized the filling station based on the precooling temperature as tabulated in Table 3.1. For nominal

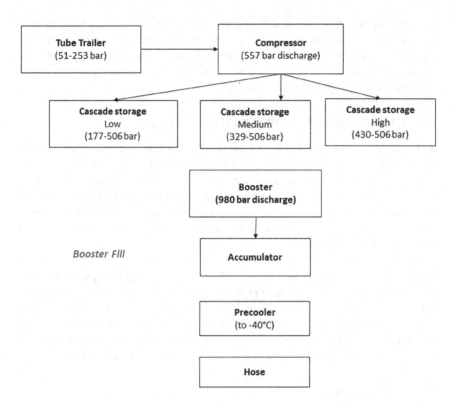

FIGURE 3.2 Booster-type refueling station for compressed hydrogen [7].

working pressure of 35 to 70 MPa with booster compressor precooling is preferred for filling stations. The precooling unit may increase the energy consumption and cost of the filling station but it may lower the temperature of compressed gas inside the tank to some extent so that it cannot go beyond the limit defined by SAE J2601 refueling protocol based on revised version 2013 [9].

SAE has also categorized the refueling stations based on the interaction or monitoring information between station and vehicle tank. The interaction between station and vehicle tank is based on refueling parameters such as temperature,

TABLE 3.1

Precooling Categories for Filling Stations [10]

Type	Temperature Range
A	Dispenser has **–40°C** precooling.
B	Dispenser has **–20°C** precooling.
C	Dispenser has **0°C** precooling.
D	Dispenser has **"No"** precooling.

pressure, density, filling rate and filling time. When station interface receives only vehicle tank pressure as a refueling parameter than station, it is called a non-communication type. However, the exchange of temperature, pressure and density takes place by the electrical interface between station and vehicle tank. It is called a communication type refueling station [11].

3.2 COMPONENTS OF REFUELING STATIONS

Hydrogen refueling station has a complex architecture since it must include high-capacity components for the distribution of hydrogen. It mainly includes a storage, compressor and refrigeration unit with additional devices essential to deliver the compressed hydrogen such as an accumulator and dispenser. The SAE developed the protocols SAE J2799 and SAE J2601 that define refueling station components, safety limits and refueling performance requirements for gaseous hydrogen refueling [11].

Figure 3.3 illustrates the configuration of a hydrogen refueling station for 70 MPa compressed hydrogen storage in fuel cell vehicles [12]. Hydrogen is supplied to the filling station through pipelines or tube trailers. The supply pressure through a pipeline is up to 20 bar and it is up to 200 bar for a tube trailer. The onsite production of hydrogen using SMR or electrolysis can generate hydrogen at a pressure of 20 bar. Regardless of the source of hydrogen production, supplied hydrogen needs to be stored at high pressure for a longer period of time.

3.2.1 STORAGE SYSTEM

Hydrogen is supplied from the source stored at the refueling site as buffer storage. Figure 3.4 presents two configurations for storage where hydrogen is pressurized to 950 bar and stored at high-pressure storage. In another configuration, hydrogen is compressed to 500 bar and stored at a medium pressure buffer storage

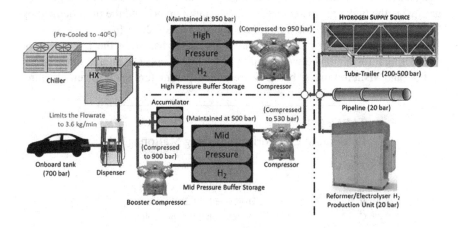

FIGURE 3.3   70 MPa Compressed hydrogen refueling station [12].

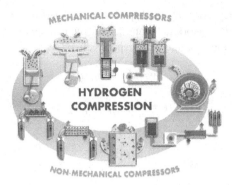

FIGURE 3.4 Hydrogen compression technologies for stationary and automotive applications [14].

system (MPS). In MPS, hydrogen is again compressed to the 900 bar by the booster compressor before dispensing to the onboard storage.

Generally, a refueling station configured for 70 MPa onboard storage stacked in high-, medium- and low-pressure storage at the refueling site is called as a cascade storage system. [13]. In high-pressure storage, six Type II tanks are used to store hydrogen at 900 bar which can store 90 kg of hydrogen. Six and eleven tanks are used to store hydrogen at a pressure of 415 bar and 200 bar, respectively in medium- and low-pressure buffer storage. The amount of hydrogen stored in low- and medium-pressure storage is 190 kg and 85 kg, respectively at full pressure. The filling process starts with low-pressure storage. If the pressure difference between storage and vehicle tank is low then it is switched to medium storage, consecutively. The major advantage of using a cascade system is that the capacity of the storage system at the refueling station can be better utilized.

3.2.2 COMPRESSOR AND BOOSTER COMPRESSOR

The primary function of the compressor is to pressurize the low-pressure hydrogen supplied from the source of production. The lower molecular weight of hydrogen restricts the use of a centrifugal compressor commonly used for gas compression. For this, a mechanical or piston compressor and diaphragm compressor are preferred for pressurizing the hydrogen at high pressure up to 900 bar with higher volumetric efficiency.

A single-stage hydraulically operated piston compressor is generally used at refueling stations to compress up to 850 bar of discharge pressure as shown in Figure 3.4 [14]. With another type of compressor called the diaphragm compressor, higher compression ratio and discharge rates can be obtained (Figure 3.4). The compressed gas obtained with this compressor is free of contamination and compression is more isentropic.

The sizing of compressor for refueling station depends on the capacity of the compressor which is defined in terms of swept volume, the number of strokes and volumetric efficiency given by the Equation 3.1. Performance of the compressor is measured in terms of work done and isentropic efficiency of compressor given by the below Equations 3.2 and 3.3, respectively:

$$w = \left(h_{out} - h_{in}\right)q \tag{3.1}$$

$$q = NV_{swept}\rho_{in}\eta_v)q \tag{3.2}$$

$$\eta_i = k\left(\frac{P_{out}}{P_{in}}\right)^n \tag{3.3}$$

where w is work done by the compressor, h stands for enthalpy of gas at inlet and outlet of the compressor, q is mass flow rate, N denotes the number of cylinders, V is swept volume, n is the speed of piston, η is efficiency and $\frac{P_{out}}{P_{in}}$ is compression ratio or pressure ratio.

3.2.3 REFRIGERATION UNIT

The compressed hydrogen passes through the variable area valves and pipelines resulting in the rise of temperature of dispensed hydrogen. The negative Joule–Thomson Coefficient of hydrogen is responsible for the increase in the temperature of hydrogen dispensed to the vehicle tank. Therefore, a refrigeration unit is required for extracting the heat of compressed hydrogen after coming out of the compressor. The refrigeration unit mainly consists of a heat exchanger and a chiller as shown in Figure 3.5.

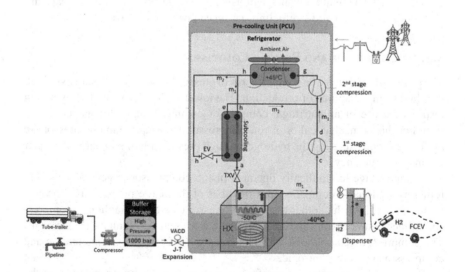

FIGURE 3.5 Precooling unit of hydrogen refueling station [15].

The precooled hydrogen from the refrigeration unit supplied to the vehicle results in lower-end temperature of the vehicle tank and higher storage density. The temperature obtained during precooling of hydrogen can be lowered up to −33°C. Precooling unit mainly includes a chiller, heat exchanger, condenser and refrigeration fluid or refrigerant. The refrigerant flows in the refrigeration cycle as in Figure 3.3 through compressor, condenser and expansion valves [15]. The portion of refrigerant from the condenser is passed to the sub-cooler to maximize the refrigeration effect [15]. The dispensed hydrogen is cooled by releasing the energy to evaporator. The temperature of dispensed hydrogen lowers up to −33°C depending upon the heat exchanger and chiller capacity. The precooling unit increases the energy requirement and initial cost of investment of refueling station due to the presence of chiller and heat exchanger.

The performance of the unit is measured by the heat extracted, effectiveness of heat exchanger and coefficient of performance of chiller given by the Equations 3.4–3.6.

$$Q = \left(h_{out} - h_{in}\right)q \tag{3.4}$$

$$\in = \frac{Actual\ heat\ transfer}{Ideal\ heat\ transfer} \tag{3.5}$$

$$COP = \frac{Q}{W} \tag{3.6}$$

where Q is the heat extracted from compressed hydrogen, $\in$ is the effectiveness of the heat exchanger, COP stands for coefficient of performance.

Apart from these thermodynamic equations, the electrical energy consumed by the refrigeration unit is also an essential parameter for measuring performance [15,16]. Because energy consumed or cooling power demand to attain or maintain the temperature in the range −40°C to −33°C is an energy-intensive process and depends upon the utilization of refueling stations [16].

3.3 ECONOMICAL ASPECTS OF REFUELING STATIONS

The total cost of the refueling station includes the sum of individual component costs. The capital investment in refueling stations depends on component sizing and hydrogen demand for particular locations. Similarly, the cost of hydrogen dispensed is also a function of individual component cost contribution called as levelized cost [17,18]. Therefore, levelized cost of dispensing hydrogen also includes production, delivery and refueling station costs. The cost of the refrigeration system, chiller and compressor varies with its capacity (Figure 3.6). Hence cost increases as capacity increases per tons or per kW as shown in Figure 3.7.

Figure 3.7 shows the breakdown of components cost and its contribution to the levelized cost of hydrogen. Figure 3.7 illustrates the contribution of compressor, storage, refrigeration unit and dispenser on levelized cost of hydrogen. The

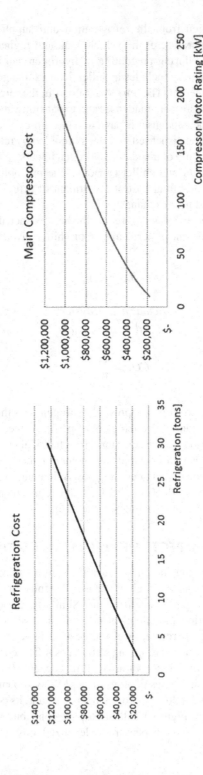

FIGURE 3.6 Cost of refrigeration and compressor in refueling stations [17].

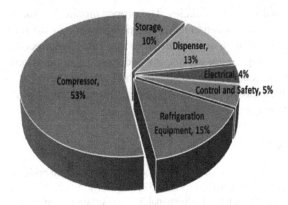

FIGURE 3.7 Cost contribution of refueling station components [17].

compression of hydrogen at high pressure makes compression as a highest contributor with 53% in total cost. Energy-intensive cooling up to the critical temperature of hydrogen makes refrigeration unit another major contributor to levelized cost of hydrogen.

Apart from these, utilization of refueling stations i.e. per day consumption or the number of vehicle fill per day also influences the cost of hydrogen dispensing. The lack of penetration of fuel cell vehicles in the market, and overutilized stations are also responsible for increasing the cost of hydrogen dispensed as back-to-back fill increases the load on refrigeration units [15,19]. The cost reduction can be achieved by increasing the number of deployed stations and greater market penetration of fuel cell vehicles.

3.4 CODES AND STANDARDS FOR REFUELING STATIONS

The purpose of codes and standards is technology validation. Competing with conventional refueling facilities and providing required safety to the customer is the ultimate goal of standards. SAE has developed Global Technical Regulation (GTR)-2004 for fuel cell vehicles refueling process and stations to achieve the desired targets. For this, SAE J2601 provides the guidelines for the refueling process of hydrogen storage systems operating at the pressure of 35–70 MPa [10,20]. The SAE J2601 also covers the refueling station and its specific configuration and dispenser type. Similarly, SAE J2799 and J2600 define the hardware/software requirements and design and testing of fittings/connections for refueling, respectively [21,22].

3.5 SUMMARY

The refueling station plays a significant role in the success of fuel cell vehicles and the performance of the refueling process. For an inherent safer refueling process and short filling time, refueling stations need to be developed to compete with conventional gasoline vehicles. The proper monitoring and control of various refueling parameters can enhance the performance of the refueling process. The high-pressure storage at the station, energy-intensive processes in refrigeration units and high-pressure compressor increase the initial investment and cost of refueling. Cost reduction is one of the challenges for refueling stations as the acceptability of hydrogen fuel cell vehicles also depends on the availability of economical hydrogen.

REFERENCES

1. Fichtner, M., *Hydrogen storage, in hydrogen economy*, Cambridge University Press, pp. 309–322, 2009. (ISBN-13 978-0-521-88216-3).
2. Faulin, J., et al., Chapter 1 - Sustainable transportation: concepts and current practices. In *Sustainable transportation and smart logistics*, J. Faulin, et al., Editors. Elsevier, Amsterdam, The Netherlands. pp. 3–23, 2019. (ISBN: 9780128142431)
3. Elgowainy, A., M. Wang, Well-to-wheel analysis of sustainable vehicle fuels. In *Encyclopedia of sustainability, science and technology*, R.A. Meyers, Editor. Springer, 2011.
4. Steffen Mausa, J.H., Chakkrit Na Ranongb, Erwin Wü¨chnera, Gerardo Friedlmeiera, David Wengera, Filling procedure for vehicles with compressed hydrogen tanks. *Internation Journal of Hydrogen Energy*, 33, 4612–4621, 2008.
5. Trill Rolf. Gaseous hydrogen refuelling stations (European Industrial Gases Association Symposium Strasbourg), 2004.
6. Schneider Jesse: Optimizing vehicle hydrogen fueling. In *Proceedings of the sixteenth national hydrogen association annual conference*, Washington, DC, 2005.
7. Pratt, J., D. Terlip, C. Ainscough, J. Kurtz, A. Elgowainy, H2 first reference station design task. NREL/SNL 2015. (https://www.nrel.gov/docs/fy15osti/64107.pdf) Accessed on June 2021.
8. Mayer, T., M. Semmel, M.A.G. Morales, K.M. Schmidt, A. Bauer, J. Wind, Techno-economic evaluation of hydrogen refueling stations with liquid or gaseous stored hydrogen. *International Journal of Hydrogen Energy*, 44(47), 25809–25833, 2019.
9. Elgowainy, A., K. Reddi, D.Y. Lee, N. Rustagi, E. Gupta, Techno-economic and thermodynamic analysis of pre-cooling systems at gaseous hydrogen refueling stations. *International Journal of Hydrogen Energy*, 42(49), 29067–29079, 2017.
10. Schneider, J., SAE J2601- Worldwide hydrogen fueling protocol: Status, standardization & implementation. SAE International, 2009.
11. Committee, F.C.S., Fueling protocols for light duty gaseous hydrogen surface vehicles. In *SAE International*. 2010, SAE International. (https://www.sae.org/standards/content/j2601_201003/) Accessed on May 2020.
12. Reddi, K., M. Mintz, A. Elgowainy, E. Sutherland, Challenges and opportunities of hydrogen delivery via pipeline, tube-trailer, liquid tanker and methanation-natural gas grid. In D. Stolten, B. Emonts, Editors. *Hydrogen science and engineering: Materials, processes, systems and technology*, vol. 1. Wiley-VCH Verlag, 2016.
13. Hydrogen and fuel cell NREL "Hydrogen infrastructure testing and research facility". https://www.nrel.gov/hydrogen/hitrf-animation.html Accessed on June 2021. Accessed on May 2021.
14. Leon, D.A., *Green energy and technoloygy (Hydrogen technology for mobile application)*. Springer-Verlag, Berlin, Heidelberg, 2008. (e-ISBN: 978-3-540-69925-5).

15. Elgowainy, A., K. Reddi, D.Y. Lee, N. Rustogi, E. Gupta, Techno-economic and thermodynamic analysis of precooling system at gaseous hydrogen refueling station. *International Journal of Hydrogen Energy*, 44, 29067–29079, 2017.
16. Diaz, C.B., Techno-economic modeling and analysis of hydrogen refueling station. *International Journal of Hydrogen Energy*, 44, 495–510, 2019.
17. Reddi, K., E. Amgad, N. Rustagi, E. Gupta, Impact of hydrogen SAE J2601 fueling methods on fueling time of light-duty fuel cell electric vehicles. *International Journal of Hydrogen Energy*, 42(26), 16675–16685, 2017.
18. Reddi, K., E. Amgad, N. Rustagi, E. Gupta, Two-tier pressure consolidation operation method for hydrogen refueling station cost reduction. *International Journal of Hydrogen Energy*, 43(26), 2919–2929, 2018.
19. Mayyas, A., M. Mann Manufacturing competitiveness analysis for hydrogen refueling stations. *International Journal of Hydrogen Energy*, 44(18), 9121–9142, 2019.
20. Buttner, William et al. Hydrogen monitoring requirements in the global technical regulation on hydrogen and fuel cell vehicles 2015. (https://www.nrel.gov/docs/fy16osti/64074.pdf) Accessed on May 2021.
21. SAE J2799 Hydrogen surface vehicle to station communications hardware and software. https://www.sae.org/standards/content/j2799_201404/ Accessed on June 2021.
22. SAE J2600 Compressed hydrogen surface vehicle fueling connection devices https://www.sae.org/standards/content/j2600_201211/ Accessed on June 2021.

4 Refueling of Compressed Hydrogen

4.1 INTRODUCTION

Refueling of compressed hydrogen in automotive vehicles run by fuel cells is a complex phenomenon by virtue of hydrogen gas's physical response. The high-pressure hydrogen can be filled from the refueling station to the vehicle tank by only connecting two storage systems, one is at high pressure while another is at low pressure. However, it becomes complicated when hydrogen changes its characteristics with every change in pressure and temperature during the flow from the refueling station to the vehicle tank. Therefore, refueling possesses two major concerns – the first one is the behavior of compressed hydrogen and the second one is fast refueling for shorter filling time to compete with conventional liquid fuel refueling.

The former one depends on the characteristics of hydrogen at different temperatures and pressures during the refueling process. Hydrogen is stored in a refueling station reservoir at a pressure of nearly 1.25 times of normal working pressure, e.g. for 70 MPa nominal working pressure, the station reservoir pressure would be 87.5 MPa [1]. When hydrogen is filled in a storage tank of fuel cell vehicles through the nozzle, it undergoes a throttling process and expands in the tank. The extent and direction of change in temperature during the constant enthalpy process is defined by the Joule–Thomson coefficient. A positive Joule–Thomson coefficient means that decrement in temperature happens along with decrement in isenthalpic pressure, while a negative Joule–Thomson coefficient means that an increment in temperature occurs along with the isenthalpic pressure decrement. Hydrogen is one of the gases having a negative Joule–Thomson coefficient. Hence, the temperature of hydrogen rises rapidly as pressure lowers down, inside the tank leading to a decrease in the density of the gas [2]. This effect may also be observed in between the station tank to the dispensing unit of hydrogen wherever restriction in flow takes place.

The latter (second) concern is related to the embracement of hydrogen as fuel in the vehicles run by fuel cell in a market where conventional vehicles dominate. The electric vehicle, which is considered to be an immediate replacement for gasoline vehicles, is very much lacking in its charging time compared to conventional vehicles [3]. However, demonstrated hydrogen fuel cell vehicles by Toyota, Honda, etc. have charging time or filling time of less than 5 minutes [4]. In order to complete with conventional vehicles and achieve the goal fixed by the United States Department of Energy (US DOE), filling time of hydrogen fuel cell should be less than 3 minutes [5]. For this, fast refueling technique is required for

DOI: 10.1201/9781003244318-4

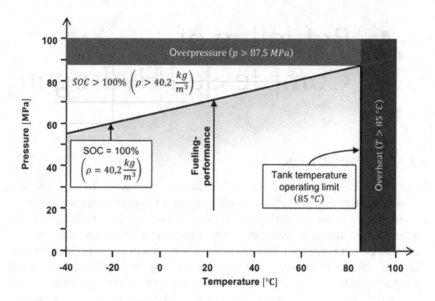

FIGURE 4.1   Hydrogen refueling window for 70 MPa compressed hydrogen tank [7].

shortening the filling time of the vehicles driven by fuel cell, simultaneously maintaining the safer refueling experience for the user, same as conventional fuels.

4.2   REFUELING PROTOCOL

The refueling of hydrogen fuel cell vehicles should offer the same comfort and safety to the user as provided by liquid fuels. The expected refueling time of less than 3 minutes can only be achieved by the fast refueling of the storage system. According to the refueling protocol defined by SAE J2601 released version 2013 hydrogen should be refueled quickly to the required density (no overfilling) while maintaining the internal tank temperature lower than 85°C (no overheating) [6,7]. Figure 4.1 shows an optimal refueling window for compressed hydrogen filling at a nominal working pressure of 70 MPa.

The SAE TIR J2601 establishes the fueling protocol guideline for onboard hydrogen refueling for a nominal working pressure of 35 MPa and 70 MPa [8,6]. The SAE TIR J2601 elaborates the technical goal of achieving the desired state of charge (SOC) of between 90–100% while maintaining the pressure and temperature below the 87.5 MPa and 85°C, respectively.

4.3   REFUELING PROCEDURE

The refueling of hydrogen in the vehicle tank depends on refueling parameters which are categorized as station and vehicle tank parameters. These parameters include pressure/temperature of refueling station storage and vehicle tank, station hydrogen delivery temperature, filling rate, vehicle tank volume, end temperature/

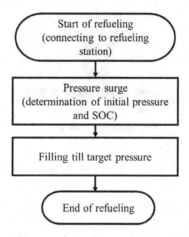

FIGURE 4.2 Flow chart of refueling steps with compressed hydrogen.

pressure of refueling and filling time. Hydrogen is filled in a vehicle tank using two approaches – Non-communication filling and communication filling.

In non-communication filling only the pressure of the vehicle tank is monitored by the refueling station control system. The compressed hydrogen is filled in the vehicle tank by the step, as shown in Figure 4.2, known as pressure-based filling. In this process, the initial pressure of the vehicle tank is determined by pressure surge rather than filling of the tank until the target pressure is achieved.

The pressure of the vehicle tank is the only deciding factor for the completion of the refueling process hence it is called non-communication type refueling [8,11]. In this process, the pressure is continuously monitored and it is ensured that the pressure increase rate (pressure ramp rate) should be within the pressure corridor defined by SAE J2601, as shown in Figure 4.3 [9].

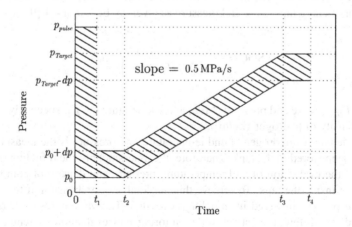

FIGURE 4.3 Pressure corridor for 70 MPa refueling [9].

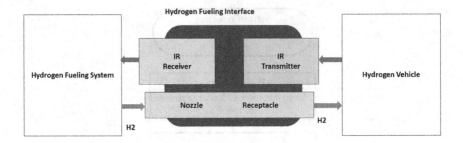

FIGURE 4.4   Refueling control in communication type station [6].

In this type of filling, the target pressure is set in such a way that the final temperature of filling remains lower than 85°C, so that a higher SOC can be obtained. But, in non-communication filling, the internal temperature of the tank cannot be monitored by the station and therefore, estimation of exact density is extremely difficult to measure. However, in communication filling, fueling information of refueling parameters such as pressure, temperature and density can be shared continuously between the station and the tank of the vehicle [10]. The communication filling approach is based on Infrared Data communication between the tank of the vehicle as shown in Figure 4.4 [6]. This approach provides transparency to the customer with more than 10% filling density. The only limitation with communication type station is the lack of flow control system and complexity with the balance of plant system components in refueling station.

4.4   ASSESSMENT OF REFUELING PROCESS

The assessment of the refueling process is only possible with the evaluation of refueling parameters such as temperature, pressure and density during the refueling. The outcome of refueling is defined as the state of charge (SOC) which is the function of end temperature and pressure as given by Equation 4.1 [9].

$$\text{SOC} = \frac{\rho_{H2}\left(P_{H2}, T_{H2}\right)}{\rho_{H2}\left(70\,MPa, 15°C\right)} \times 100 \qquad (4.1)$$

Where P_{H2}, T_{H2} are end pressure (MPa) and temperature (°C), respectively and ρ_{H2} is the density of hydrogen (kg/m³).

The density of hydrogen at end temperature and pressure is the measurement of hydrogen stored in the tank. Therefore, the assessment of the refueling process involves the evaluation of end temperature, pressure and density of compressed hydrogen inside the tank. To achieve this, in-depth comprehension of the critical physical process involved in refueling is needed. The various researchers have reported the different practices for monitoring and evaluation of temperature, pressure and density during the refueling process. The best practices include experimental studies and analytical modeling of the most recent computational fluid dynamic (CFD) simulation.

A very few experimental studies have been performed and available in the literature due to the lack of standards, technical regulations and safety issues. The Joint Research Centre of European Commission and Institute for Energy and Transport (JRC IET) Netherlands, National Research Council of Canada's Institute for Fuel Cell Innovation and Air Liquid France have performed the experimental part and related analysis of the refueling process.

The JRC IET at their reference laboratory known as Gas Testing Facility (GasTeF) has been designed to test the Type III and IV tank at high pressure [11]. The GasTeF is composed of a half-buried bunker (made of concrete) and a hydrogen storage tank. The storage tank is placed inside the sleeve that filled the inert mediums (nitrogen or helium) used for detecting permeation from the test tank as shown in Figure 4.5.

The fast-filling experiments are performed on a 29 L tank Type IV tank at the nominal working pressure of 70 MPa. The test tank is attached with thermocouples located near to the inlet, at the centerline of the tank, in the proximity of the wall and the outer surface of the tank, as depicted in Figure 4.6. For measuring the pressure, transducers are used at the inlet, outlet and inside of the tank. The GasTeF has performed a considerable number of experiments for measuring the evolution of temperature and pressure in Type IV tanks. The GasTeF JRC database contains the measured temperature and pressure for many filling conditions such as different supply temperatures, initial pressure and temperature, and density [12].

The Air Liquide company at their filling facility tested a 90.5 L Type IV tank with the thermoplastic liner at a nominal working pressure of 70 MPa [14]. The test tank is connected with thermocouples located mainly at the centerline and outer surface of the tank (Figure 4.7). The horizontally placed tank filled at a pressure ramp rate of 0.007–0.17 MPa/s for 200 seconds. The experiments were performed for measuring the temperature rise rates during the filling. They performed eight tests at different pressure ramp rates to evaluate the average temperature rise

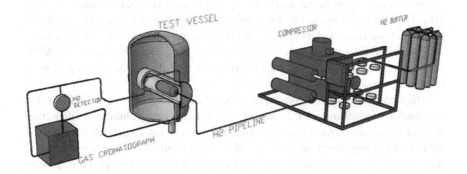

FIGURE 4.5 Schematic of GasTeF at the Joint Research Centre Netherlands [11].

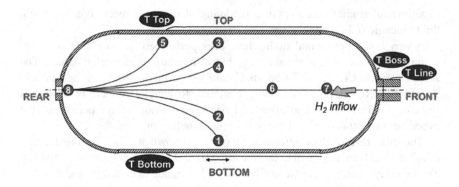

FIGURE 4.6   Arrangement of thermocouples in the tank [13].

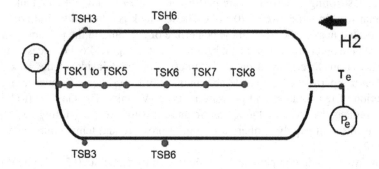

FIGURE 4.7   Arrangements of thermocouples in filling facility at Air Liquid [15].

when the filling was going on. In these experiments, thermocouples were placed away from the tank inlet, the effect of sudden expansion at the inlet has been neglected.

The National Research Council of Canada's Institute for Fuel Cell Innovation has performed experimental work on Type III and IV Dynetek cylinders of 74 L [16]. They arranged the 63 thermocouples and two pressure transducers inside the tank which measure the temperature and pressure at various points inside the tank. The tank was filled at the starting temperature of 20°C and pressure of 9.36 MPa, and the mean temperature and temperature profile inside the tank were measured.

All three experimental practices mentioned above have almost the same approaches of inserting the thermocouples inside the tank volume but differ from the tank size. A large number of thermocouples can be inserted in test tanks which measure the temperature inequality such as stratification in large-size tanks and horizontal temperature gradient. The experimental investigation provides precise measurement of temperature but it may vary due to different levels of accuracy of measuring instruments.

Another approach for evaluation of the refueling process adopted by many researchers was the analytical modeling of the process. The analytical models refer to equations set on scientific rules solved for the refueling parameters. Before developing a mathematical model, it is essential to understand the various physical phenomena involved in the process. Many of the authors reported their model based on basic thermodynamics laws which include conservation of mass and energy for defining the filling of the tank [17–19]. Most of them have presented a time-based 3-dimensional model of these laws and applied them to the control volume of the tank (given by the Equations 4.2 and 4.3) for estimation of temperature and pressure in the required filling time [19–21]. The balancing of equations derives the temperature and pressure of hydrogen inside the control volume.

Conservation of mass

$$q = \frac{dm_{H2}}{dt} \tag{4.2}$$

Conservation of energy

$$Q - w = \frac{\partial}{\partial x}\left[\int_{CV}\left(u + \frac{1}{2}v^2 + gZ\right)\rho dV\right] + \int_{CS}\left(h + \frac{1}{2}v^2 + gZ\right)\rho dA) \tag{4.3}$$

Where q is flow rate, m_{H2} is mass of hydrogen, Q is heat flux, W is viscous work in the control volume, u is internal energy, v is the velocity at the inlet, g is gravity, Z is the potential difference, ρ is density, V is control volume, h is enthalpy.

In order to obtain an exact estimation of density at particular pressure and temperature, some of the researchers also reported the applications of the equation of state to control volume [22,23]. Due to the significant effect of compressibility at different pressures and temperatures hydrogen deviates from ideal gas behavior. Therefore, the amount of hydrogen filled inside the control volume has been defined by using the equation of state for real gases. In general, the reported equations of state for estimation of compressed hydrogen density in control volume are Van der Waals (VW), Redlich–Kwong (RK), Soave–Redlich–Kwong (SRK) and Peng–Robinson (PR) (given by the Equations 4.4–4.7) [23].

Van der Waals

$$\left(P + \frac{a}{V_m^2}\right)(V_m - b) = RT \tag{4.4}$$

Redlich–Kwong

$$P = \frac{RT}{(V_m - b)} - \frac{a}{\sqrt{T}V_m(V_m + b)} \tag{4.5}$$

Soave–Redlich–Kwong (modification of Redlich–Kwong)

$$P = \frac{RT}{\left(V_m - b\right)} - \frac{a\alpha}{V_m\left(V_m + b\right)} \tag{4.6}$$

Peng–Robinson

$$P = \frac{RT}{\left(V_m - b\right)} - \frac{a\alpha}{\left(V_m^2 + 2bV_m + b^2\right)} \tag{4.7}$$

Where P, R, V_m and T are pressure, gas constant, molar volume and temperature, respectively. a and b are specific constants depending on the critical properties of the gas. The α in Equations 4.6 and 4.7 is a function of vapor pressure and temperature.

4.5 REFUELING AT CONSTANT FILLING RATE

The compressed hydrogen is filled in a fuel cell vehicle using a pressure-based filling approach. In this approach, the vehicle tank is filled up to the target pressure at a defined pressure ramp rate. But due to the physical behavior of hydrogen, it cannot be filled up to the desired SOC, in a shorter time. To achieve this, the vehicle tank can be filled at a constant filling rate and attend the desired SOC of more than 90%.

4.5.1 Case Study: Fast Filling Simulation of 29 L Type IV tank [24]

For this, refueling process simulation of 29 L Type IV tank was conducted to fill the tank at the constant filling rate of 2 g/s, 4 g/s, 6 g/s, 8 g/s and 10 g/s for 200 seconds, at supply temperatures of −40°C, −20°C, 0°C and 15°C (as in Figure 4.8) defined by SAE J2601 using CFD approach [24].

The CFD simulation has considered the refueling station and vehicle tank parameters to simulate the refueling process. The supply temperature, rate of filling and initial temperature/pressure are essential station and vehicle tank

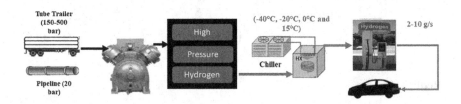

FIGURE 4.8   Schematic of refueling process.

parameters, respectively considered for refueling simulation. Type IV tank is a tank made of composite materials having low thermal conductivity; therefore, refueling simulation was considered as an adiabatic refueling simulation. The refueling process is evaluated for temperature, pressure and density attained at the end of refueling. Based on this, SOC has been calculated for all filling rates and supply temperatures.

Increment of hydrogen temperature inside the tank during filling at different supply hydrogen temperatures is shown in Figures 4.9 and 4.10. It should be noted that the sharp temperature rise was observed at the initial stage of filling for 50–60 seconds, which is mainly due to the sudden expansion of compressed gas from the injector to the tank. The real behavior of hydrogen and the negative Joule–Thomson coefficient lead to a high-temperature rise rate at the initial stage of filling. Beyond this stage, the temperature rise was moderate until the end of refueling. The mass of hydrogen inside the tank increases as filling proceeds. In the later stage, low temperature rise rate is observed as the temperature is inversely proportional to the mass content.

For all supply hydrogen temperatures, a similar trend in temperature rise has been observed. However, the temperature rise rate was higher after 60 seconds of filling for supply temperature of 15°C and 0°C (Figure 4.9(a) and (b)) compared to the filling at −20°C and −40°C, as shown in Figure 4.10(a) and (b). A similar observation was reported by Melideo et al. and Galassi et al. in their study of compressed gas refueling in Type III and IV tanks, respectively [25,26].

The data for all four supply hydrogen temperatures show a similar temperature rise pattern, however, the temperature attained at the end of refueling is different and depends on the filling rates. The higher turbulence and entropy lead to an increase in temperature at high refueling rates. At 15°C, the end temperature attained lies beyond 85°C, the limit defined by SAE J2601 for filling rates of 8 g/s and 10 g/s. As supply temperature reduces to 0°C the end temperature lies within the limit even at higher filling rates. This encouraged the supply of hydrogen at

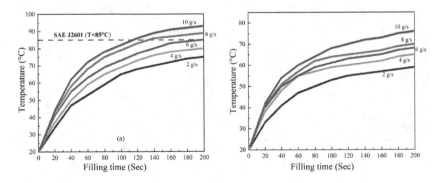

FIGURE 4.9 Evolution of temperature of hydrogen inside the tank during refueling at different filling rates and supply temperatures of (a) 15°C (b) 0°C.

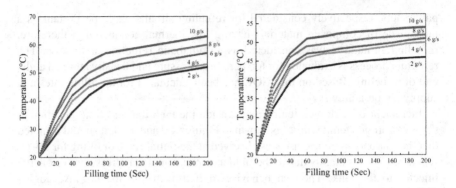

FIGURE 4.10 Evolution of temperature of hydrogen inside the tank during refueling at different filling rates and supply temperatures of (a) −20°C (b) −40°C.

low temperature or precooled temperature. The end temperature attained has shown a significant difference at the supply temperatures of −20°C and −40°C for different filling rates (Figure 4.10(a) and (b)).

Figure 4.11 shows the change in hydrogen pressure inside the tank during refueling at different filling rates and supply temperatures. A linear increment of

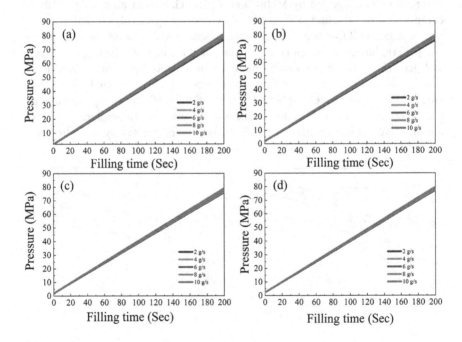

FIGURE 4.11 Increment of hydrogen pressure of the tank during refueling at different filling rates and supply temperatures of (a) 15°C (b) 0°C (c) −20°C (d) −40°C.

pressure from an initial pressure of 2 MPa was observed at all supply temperatures and filling rates. As filling proceeds, pressure increases gradually due to the compression of hydrogen inside the tank. The supply temperature did not contribute much to the development of pressure inside the tank due to the adiabatic expansion of gas. However, the difference of 2–3 MPa in end pressure was observed for lower to higher filling rates.

Figure 4.12 represents the evolution of the density of compressed hydrogen inside the tank during the refueling. It was observed that the nature of the curve does not match with the asymptotic and linear pattern of growth of temperature and pressure, respectively. However, the density of compressed hydrogen was greatly affected by the growing temperature and pressure within the tank due to filling rates and supply temperature.

For the supply temperature of 15°C and 0°C, the density data at the end of refueling were varied between 34–36 kg/m^3 and 35–37 $kg/m,^3$ respectively, for different filling rates. Similarly, density lies in the range of 36–38 kg/m^3 and 38–39 kg/m^3 for the supply temperatures of −20°C and −40°C, respectively. The density values achieved at −20°C and −40°C were found closer to the target density of 40.2 kg/m^3 achieved at nominal working conditions of 70 MPa and 15°C.

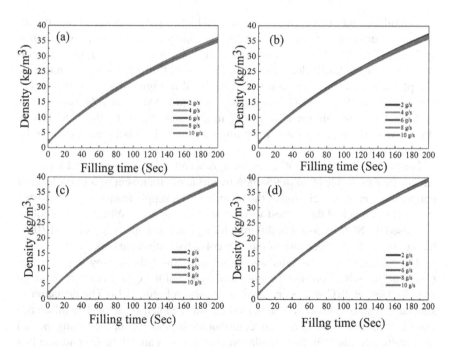

FIGURE 4.12 Evolution of density of hydrogen inside the tank during refueling at different filling rates and supply temperatures of (a) 15°C (b) 0°C (c) −20°C (d) −40°C.

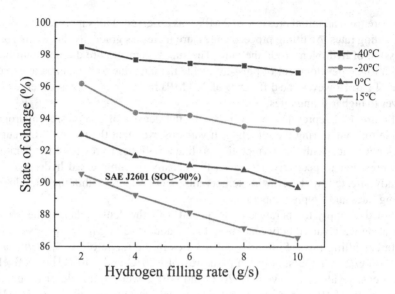

FIGURE 4.13   State of charge at the end of refueling for different supply temperatures and filling rates of hydrogen.

The fulfillment of refueling was analyzed by the SOC obtained at the conclusion of refueling at different filling rates and supply temperatures as represented by Figure 4.13. The SOC obtained has been influenced by the supply temperature and filling rates. It decreases with increasing filling rates and attained the acceptable peak at low filling rates, as depicted in Figure 4.13. For the supply temperature of 15°C, the SOC stays under the limiting value of 90% as described by SAE J2601 for almost all types of filling rates. For 0°C, the SOC hovers slightly over the limiting value for filling rates of 2–8 g/s but remains under the mark of 10 g/s.

For precooled H_2, the SOC remains significantly above the required level and attains the admissible level of 95–98% by virtue of attainment of a lower level of temperature on the conclusion of the refueling. At a supply temperature of −40°C, the SOC has touched the highest value of 98.48% for 2 g/s. While the precooling has raised the SOC by 3–4% at different filling rates, still the target value of 100% has not been attained because of the heat collected within the tank.

The SOC obtained by adiabatic simulation was validated by experimental results of the SOC. For this, the simulated SOC and the experimentally obtained SOC by GasTef JRC were compared, as shown in Figure 4.14. The deviation of simulated results from experimental values was denoted by error. The difference was observed due to the two major contributors responsible for increasing the end temperature during refueling simulation. The first was an adiabatic condition that did not permit any heat transfer from the tank walls. Another one was the target of achieving a short filling time which accumulates a huge quantity of heat within

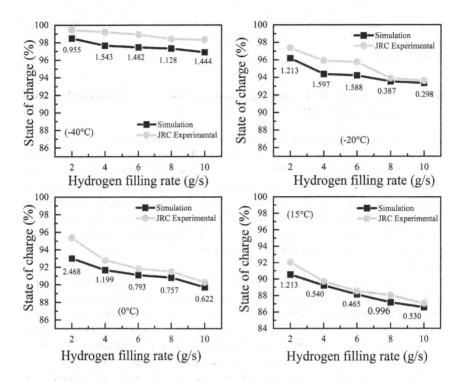

FIGURE 4.14 Comparison of simulated state of charge and experiment state of charge obtained from JRC.

the tank at fast-filling rates. In both circumstances, the end temperature of refueling increases and eventually lowers the SOC.

The small difference in the SOC also signifies that the simulation approach adopted was feasible enough for the analysis of the refueling process. It can be applied to any size of tanks for the analysis of the refueling process, irrespective of the volume.

The refueling at constant filling rates suggested that the overheating of the tank at ambient supply temperature and higher filling rates leads to a low SOC. However, filling at low and moderate rates can attain the desired SOC of more than 90% even at ambient supply temperature. With a precooled supply of hydrogen, an improved or higher SOC can be obtained. The refueling parameters significantly affect the SOC which eventually governs the driving range of fuel cell vehicles.

4.6 REFUELING AT VARIABLE FILLING RATE

The temperature and pressure rise in Type IV tank during the refueling process affect the SOC. This might be caused by a lack of flow regulation and monitoring of refueling parameters at the refueling station. Hydrogen supply temperature, filling rate, temperature/pressure at the end of refueling, density and filling time are essential refueling parameters needed to be controlled and monitored at the refueling station.

4.6.1 CASE STUDY-II: EXAMINATION OF REFUELING PARAMETERS FOR VARIABLE FILLING RATES [27]

The refueling of compressed hydrogen at a fixed filling rate indicated that the decrement in the supply temperature reduces the end temperature of refueling, even at higher filling rates. This resulted in higher SoC attained at the conclusion of the refueling. Hence, the possible role of each parameter, which may influence the storage density for various filling settings, was required to undergo examination For this, the contribution of refueling parameters on outcome was examined by the root cause analysis process (RCA), which covers the assessment of process-variables which affects the quantum of process, either explicitly or by another way.

There are three stages of RCA. The first stage is definement and categorization of variables; the second stage is the selection of the variables for regression study and the third stage is predictor standing calculated on the probability correlation factor (F-value) of variables [28]. The related program will calculate the predictor statistics followed by the prediction of rank, based on the predictor importance. Further, the importance plot developed by RCA presents the contribution of process variables in affecting the yield of the process individually and mutually by order of interactions of variables. The high order of interaction of variables in the plot shows the combined contributions of variables, which affects the outcome of the process. The hydrogen supply temperature, filling rate, and filling time are identified as directly involved variables in the refueling process.

Figure 4.15 depicts "The Importance Plot" as the relation between filling time (t), filling rate (q) and end temperature (T)/pressure (P), correlated with the Third- and Fourth-order interactions. The Importance plot reveals that the pressure (P) and filling time (t) exclusively contributed extra with a larger F-value. The Second-order interaction emphasizes that increment in pressure (P) in an association with filling rate (q), when the filling is going on, eventually enhances the storage density. Conversely, Second- and Third-order interactions of P, t, q with T reduce their contribution in the Importance Plot. Although T was found to contribute to storage density, its Third-order interaction for t and q contributed high and eventually gets upward position in the Importance Plot. The Fourth-order interaction of process parameters gets the least position rank, because of individual F-value.

The maximum F-value for end pressure (P)/temperature (T), filling rate and filling time endorses the factors as serious contributors to storage density. Almost the same effect of end temperature (T) was noticed in the constant refueling rate simulation. As well as in the previously described literature of numerical model of density evolved using a real gas equation of state where SoC is a function of

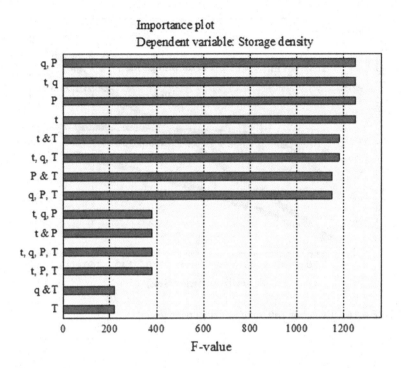

FIGURE 4.15   Contribution of refueling process parameters on storage density [28].

temperature and pressure of the gas [29,30]. The RCA has been utilized in various such works for process optimization [28,29] and evaluation of process parameters [30–32]. They reported a similar approach for evaluating the role of process parameters in optimizing the yield of the process. The identified parameters by RCA are used for achieving the variable filling rates.

4.6.2   FILLING PATTERN FOR VARIABLE FILLING RATE

The evolution of temperature, pressure and density has shown a distinguished growth inside the tank during the refueling of hydrogen as depicted in Figure 4.16. The rapid increase in temperature was observed initially up to 60 seconds, afterward, it increases almost linearly for all filling rates. Based on the evolution of refueling parameters specifically temperature, different filling profiles were investigated to achieve the suitable filling pattern for variable filling of hydrogen as in Figure 4.17. The rapid increase in filling rate from low to high for the initial 60 seconds of filling then a linear decrease up to the lowest designated as filling pattern-I. In filling pattern-II, filling started at a higher filling rate for initial 60 seconds then the filling rates dropped for the remaining part of the filling. However, a linearly decreasing filling profile was considered in filling pattern-III. All three

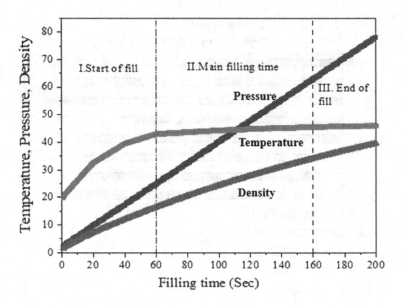

FIGURE 4.16 Evolution pattern of refueling parameters.

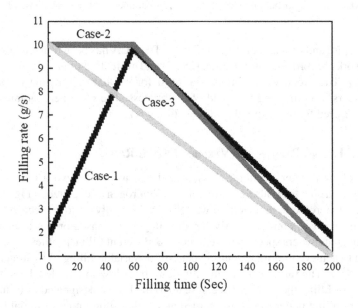

FIGURE 4.17 Refueling pattern for variable filling of compressed hydrogen.

patterns were analyzed using the multivariable regression analysis technique for predicting the temperature, pressure and density at the end of refueling.

4.6.3 Predicted Temperature, Pressure and Density

The identified refueling parameters by RCA were used to develop a regression model to predict the temperature, pressure and density. Table 4.1 presents the multivariable regression model for the end temperature of the tank. The filling time, filling rate and pressure act as independent variables for predicting the dependent variable temperature, at the end of refueling.

The predicted temperature at different hydrogen supply temperatures and filling patterns is tabulated in Table 4.2. It was observed that the predicted end temperature lies within the limiting value of 85°C explained by SAE J2601 even at higher supply temperatures. The filling pattern-II has depicted the lowest end temperature reached at the end of refueling time, for almost all of the supply temperatures. Therefore, the variable filling has taken advantage over constant refueling by restricting the end temperature to less than 85°C even at higher supply temperatures.

The pressure attained at the end of refueling with various filling patterns has been found by predicting the pressure using the regression model, as tabulated in Table 4.3. The predicted value of pressure for the entire filling time has been matched with the

TABLE 4.1
Regression Model for End Temperature at Different Supply Temperatures

Hydrogen Supply Temperature	Regression Model
−40°C	$21.83288 - (2.16047)t + (0.546)q + (5.87553)P$
−20°C	$21.59674 - (1.12452)t + (0.97063)q + (3.38074)P$
0°C	$18.39780 - (2.76512)t + (0.82143)q + (7.8630)P$
−15°C	$29.9244 + (1.3819)t + (2.8259)q + (2.97307)P$

TABLE 4.2
Predicted End Temperature for Different Filling Pattern and Supply Temperature

Supply Temperature	Filling Pattern		
	I	II	III
−40°C	51.058°C	50.319°C	53.535°C
−20°C	56.555°C	55.439°C	55.483°C
0°C	65.851°C	65.898°C	65.950°C
15°C	80.169°C	78.066°C	78.440°C

TABLE 4.3
Regression Model for Pressure at Various Supply Temperature

Hydrogen Supply Temperature	Regression Model
−40°C	$1.52279 + (0.382608)t + (0.543332)q + (0.003950)T$
−2 0°C	$1.186629 + (0.373113)t + (0.99973)q + (0.006033)T$
0°C	$1.209663 + (0.37252)t + (0.0843)q + (0.007105)T$
−15°C	$1.207916 + (0.379925)t + (0.094512)q + (0.005336)T$

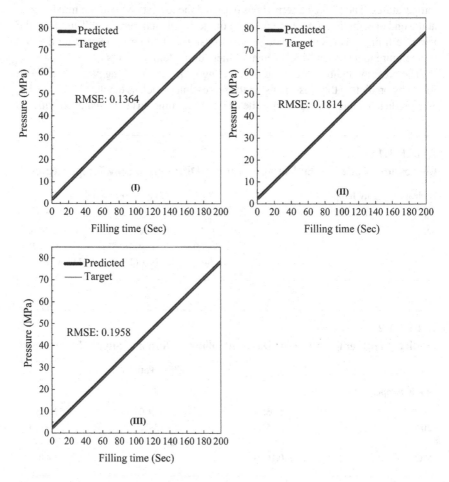

FIGURE 4.18 Predicted pressure at the end of refueling for various filling patterns at supply temperature of −40°C.

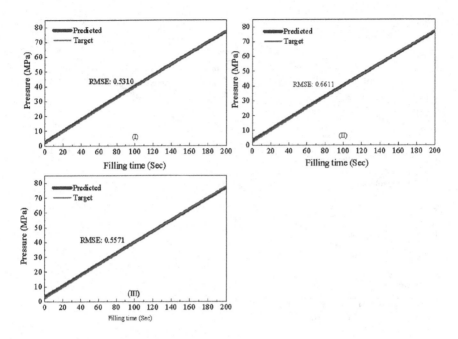

FIGURE 4.19 Predicted pressure at the end of refueling for various filling patterns at a supply temperature of −20°C.

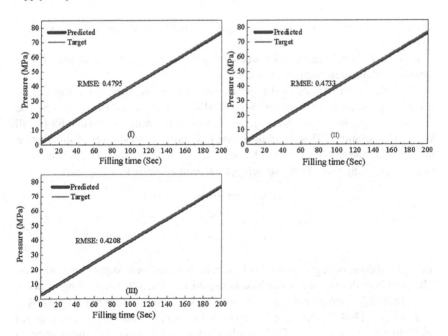

FIGURE 4.20 Predicted pressure at the end of refueling for various filling patterns at a supply temperature of 0°C.

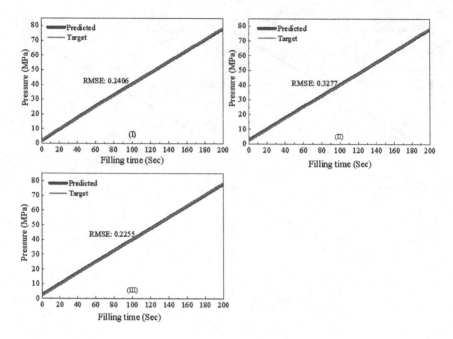

FIGURE 4.21 Predicted pressure at the end of refueling for various filling patterns at a supply temperature of 15°C.

target pressure required to achieve the target density of 40.23 kg/m³ at the pressure of 70 MPa.

Figures 4.18–4.21 present the predicted and target pressure for all three filling patterns and different supply temperatures.

The relative fit of predicted pressure with target pressure was measured by calculating the root mean square error (RMSE) given by the Equation 4.8. The deviation of predicted and target pressure lies in the range of 0.10–0.65 for all supply temperatures. This small deviation has shown a good level of satisfaction in predicting the pressure and model accuracy. This also shows decent suitability of all three filling patterns for refueling hydrogen at variable filling rates.

$$RMSE = \sqrt{\frac{\sum_{i=1}^{n}\left(P_{target} - P_{predicted}\right)^2}{n}} \tag{4.8}$$

Table 4.4 shows the regression model for density at different supply temperatures. The predicted density was examined based on the SAEJ2601 refueling protocol for permissible operating range at the pressure of 70 MPa, as shown in Figure 4.1.

Figures 4.22–4.25 represent the predicted density for all supply temperatures and filling patterns. For –40°C, the predicted density has crossed the mark of target

TABLE 4.4
Regression Model for Density at Various Supply Temperature

Hydrogen Supply Temperature	Regression Model
−40°C	$−0.00898 + (0.67809)t − (0.16149)q − (0.224222)T + (1.34277)P$
−20°C	$0.407252 + (0.497437)t + (0.223596)q − (0.941199)T + (0.248546)P$
0°C	$2.45686 + (0.89888)t + (0.18757)q − (2.03571)T + (0.17413)P$
−15°C	$2.420863 + (0.05838)t + (0.108291)q − (0.202634)T + (0.387727)P$

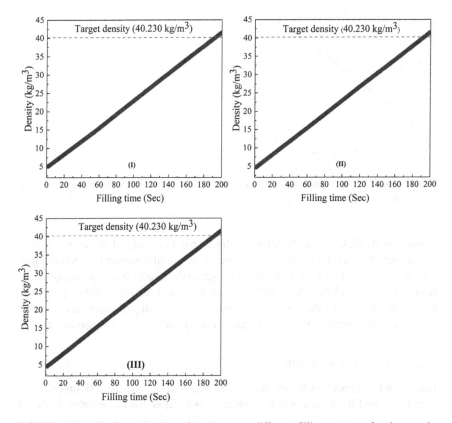

FIGURE 4.22 Predicted density of hydrogen at different filling patterns for the supply temperature of −40°C.

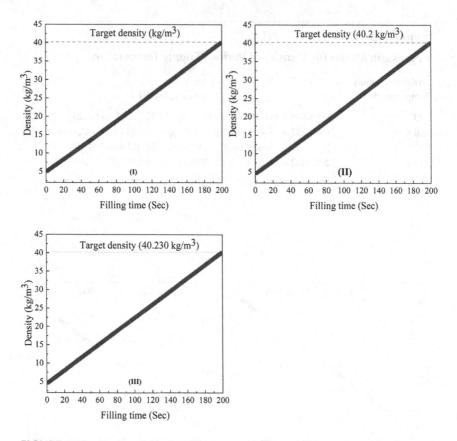

FIGURE 4.23 Predicted density of hydrogen at different filling patterns for the supply temperature of −20°C.

density of 40.23 kg/m³ at 70 MPa which has been considered as an overdensity case as per the SAE J2601 and not recommended by this protocol for refueling of fuel cell vehicles. However, it has come closer to the target density for supply temperature of −20°C. The density of 40.177 kg/m³ was observed for filling pattern-II leading to the SOC of 99.868%. Neglecting the overdensity at supply temperature of −40°C, filling pattern II has the highest density for all supply temperatures.

4.6.4 COMPARATIVE STUDY

Based on the predicted temperature, pressure and density for all supply temperatures, the filling pattern II has emerged as suitable for the variable filling of

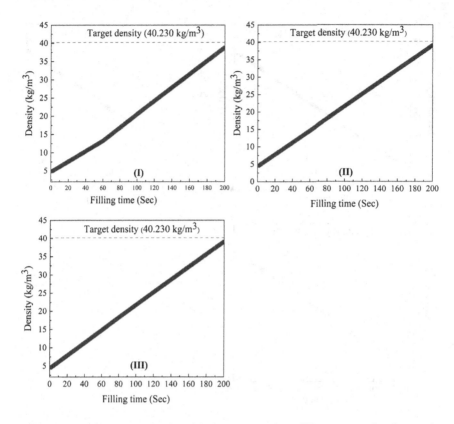

FIGURE 4.24 Predicted density of hydrogen at various filling patterns for the supply temperature of 0°C.

compressed hydrogen. The highest density, lowest temperature and acceptable range of RMSE were observed in filling pattern II. The SOC obtained from filling pattern II was compared with the best results for the SoC achieved from constant refueling and experimental studies by GasTeF JRC, as displayed in Table 4.5.

A significant improvement in the SOC was observed with variable filling using filling pattern II. The results suggest that controlling the filling rate based on vehicle tank parameters lowers the end temperature, leading to an enhanced SOC. The highest density obtained at −20°C instead of −40°C as in constant refueling also suggest that the variable filling rates reduce precooling demand of hydrogen leading to less energy consumption at the refueling station.

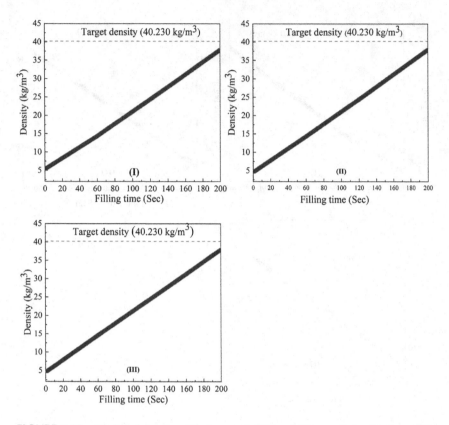

FIGURE 4.25 Predicted density of hydrogen at different filling patterns for the supply temperature of 15°C.

TABLE 4.5
Comparison of SOC

SOC of Filling Pattern II at −20°C	Highest State of Charge for Constant Refueling at −20°C	Highest SOC Achieved from Constant Refueling Process	Experimental State of Charge Achieved by GasTef JRC
99.868%	96.183%	98.483% (at supply temperature of −40°C)	**97.373%** (at supply temperature of −20°C)

4.7 SUMMARY

Refueling of compressed hydrogen is a complex phenomenon with a major concern of filling in a shorter time with the desired SOC. The constant filling of hydrogen can attain the desired storage density but gets influenced by larger supply temperatures and filling rates. Refueling parameters significantly influence the SoC attained at the end of refueling. With regulation and control of refueling parameters, the SOC can be improved. The variable filling can enhance the storage density compared to constant filling rates and also limits the precooling demand of the refueling station.

REFERENCES

1. Gardiner, M.R., J. Cunningham, R.M. Moore, Compressed hydrogen storage for fuel cell vehicles. SAE Technical Paper 2001-01-2531, 2001.
2. Bourgeois, T., F. Ammouri, D. Baraldi, P. Moretto. The temperature evolution in compressed gas filling process: A review. *International Journal of Hydrogen Energy*, 43, 2268–2292, 2018.
3. Moghbelli, H., K. Ganapavarapu, R. Langari, A comparative review of fuel cell vehicles (FCVs) and hybrid electric vehicles (HEVs) part II: control strategies, power train, total cost, infrastructure, new developments, and manufacturing & commercialization. SAE Technical Paper 2003-01-2299, 2014.
4. Hasegawa, T., H. Imanishi, M. Nada, Y. Ikogi, Development of the fuel cell system in the Mirai FCV. SAE Technical Paper 2016-01-1185, 2016.
5. Hydrogen Staorage Technologies Road Map US DOE, Editor. 2017: U.S. DRIVE (Driving Research and Innovation for Vehicle efficiency and Energy sustainability) (https://www.energy.gov/sites/prod/files/2017/08/f36/hdtt_roadmap_July2017.pdf) Accessed on January 2019.
6. Schneider, J., SAE J2601-Worldwide hydrogen fueling protocol: Status, standardization & implementation. SAE International, 2009.
7. Bauer, A., T. Mayer, M. Semmel, M.A.G. Morales, J. Wind, Energetic evaluation of hydrogen refueling stations with liquid or gaseous stored hydrogen. *International Journal of Hydrogen Energy*, 44(13), 6795–6812, 2019.
8. Buttner, W., et al., Hydrogen monitoring requirements in the global technical regulation on hydrogen and fuel cell vehicles. *International Journal of Hydrogen Energy*, 942(11), 76644-7611, 2017.
9. Steffen Mausa, J.H., S. Maus, J. Hapke, C.N. Ranong, E. Wüchner, G. Friedlmeier, D. Wenger, Filling procedure for vehicles with compressed hydrogen tanks. *International Journal of Hydrogen Energy*, 33, 4612–4621, 2008.
10. Jesse Schneider, B.R., S. Mathison Light duty fuel cell electric vehicle hydrogen fueling protocol. 2013. (https://www.energy.gov/eere/fuelcells/downloads/light-duty-fuel-cell-electric-vehicle-hydrogen-fueling-protocol). Accessed on July 2019.
11. Ortiz Cebolla, R., et al., GASTEF: The high pressure gas tank testing facility of the European commission joint research centre. *International Journal of Hydrogen Energy*, 44(16), 8601–8614, 2019.
12. Acosta, B., P. Moretto, N. de Miguel, R. Ortiz, F. Harskamp, C. Bonato, JRC reference data from experiments of on-board hydrogen tanks fast filling. *International Journal of Hydrogen Energy*, 39(35), 20531–20537, 2014.
13. Cebolla, R.O., B. Acosta, P. Moretto, N. Frischauf, F. Harskamp, C. Bonato, D. Baraldi, Hydrogen tank first filling experiments at the JRC-IET GasTeF facility. *International Journal of Hydrogen Energy*, 39(11), 6261–6267, 2014.
14. Bourgeois, T., F. Ammouri, M. Weber, C. Knapik, Evaluating the temperature inside a tank during a filling with highly- pressurized gas. *International Journal of Hydrogen Energy*, 40, 11748–11755, 2015.

15. Bourgeois, T., et al., Evaluating the temperature inside a tank during a filling with highly-pressurized gas. *International Journal of Hydrogen Energy*, 40(35), 11748–11755,2015.

16. Dicken, C.J.B., W. Mérida, Modeling the transient temperature distribution within a hydrogen cylinder during refueling. *Numerical Heat Transfer, Part A: Applications*, 53(7), 685–708, 2007.

17. Yang, J.C., A thermodynamic analysis of refueling of a hydrogen tank. *International Journal of Hydrogen Energy*, 34(16), 6712–6721, 2009.

18. Zhao, Y., et al., Numerical study on fast filling of 70 MPa type III cylinder for hydrogen vehicle. *International Journal of Hydrogen Energy*, 37(22), 17517–17522, 2012.

19. Xiao, J., P. Bénard, R. Chahine, Estimation of final hydrogen temperature from refueling parameters. *International Journal of Hydrogen Energy*, 42(11), 7521–7528, 2017.

20. Xiao, J., P. Bénard, R. Chahine, Charge-discharge cycle thermodynamics for compression hydrogen storage system. *International Journal of Hydrogen Energy*, 41(12), 5531–5539, 2016.

21. Olmos, F., V.I. Manousiouthakis, Hydrogen car fill-up process modeling and simulation. *International Journal of Hydrogen Energy*, 38(8), 3401–3418, 2013.

22. Nasrifar, K., Comparative study of eleven equations of state in predicting the thermodynamic properties of hydrogen. *International Journal of Hydrogen Energy*, 35(8), 3802–3811, 2010.

23. Bai-Gang, S., Z. Dong-Sheng, L. Fu-Shui, A new equation of state for hydrogen gas. *International Journal of Hydrogen Energy*, 37(1), 932–935, 2012.

24. Sapre, S., et al., H_2 refueling assessment of composite storage tank for fuel cell vehicle. *International Journal of Hydrogen Energy*, 44(42), 23699–23707, 2019.

25. Galassi, M.C., E. Papanikolaou, M. Heitsch, D. Baraldi, B.A. Iborra, P. Moretto, Assessment of CFD models for hydrogen fast filling simulations. *International Journal of Hydrogen Energy*, 39(11), 6252–6260, 2014.

26. Melideo, D., D. Baraldi, B. Acosta-Iborra, R.O. Cebolla, P. Moretto, CFD simulations of filling and emptying of hydrogen tanks. *International Journal of Hydrogen Energy*, 42(11), 7304–7313, 2017.

27. Sapre, S., K. Pareek, M. Vyas, Impact of refueling parameters on storage density of compressed hydrogen storage tank. *International Journal of Hydrogen Energy*. https://doi.org/10.1016/j.ijhydene.2020.08.136

28. Dassau, E., D. Lewin, Optimization-based root cause analysis. In *Computer aided chemical engineering*, W. Marquardt and C. Pantelides, Editors. Elsevier. pp. 943–948, 2006. https://doi.org/10.1016/S1570-7946(06)80167-7

29. Zheng, J., X. Zhang, P. Xu, C. Gu, B. Wu, Y. Hou, Standardized equation for hydrogen gas compressibility factor for fuel consumption applications. *International Journal of Hydrogen Energy*, 41(15), 6610–6617, 2016.

30. Emmanuel Ruffio, D., D.P. Saury, Thermodynamic analysis of hydrogen tank filling. Effects of heat losses and filling rate optimization. *International Journal of Hydrogen Energy*, 39, 12701–12714, 2014.

31. Zhong, X., et al., Root cause analysis and diagnosis of solid oxide fuel cell system oscillations based on data and topology-based model. *Applied Energy*, 267, 114968, 2020.

32. Wang, J., et al., Root-cause analysis of occurring alarms in thermal power plants based on Bayesian networks. *International Journal of Electrical Power & Energy Systems*, 103, 67–74, 2018.

5 Heat Transfer Analysis

5.1 INTRODUCTION

Heat transfer plays a significant role in any process executed at higher pressure and temperature. Hydrogen supplied to the fuel cell vehicle tank passed through many cross-sections of different diameters and lengths. Any variation in the passage of hydrogen changes its characteristics due to the variation in pressure and temperature. The heat accumulated inside the Type IV tank occurs due to two vital reasons.

The first reason corresponds to the negative Joule–Thomson coefficient of H_2. This results in increment of temperature of H_2 when it passes through restricted openings. Heat generation is observed when hydrogen is dispensed to the vehicle tank. It is important to emphasize that very restricted opening increases the temperature but continuous isenthalpic expansion is observed at the vehicle tank causing the accumulation of a huge quantity of heat within the tank.

The second reason corresponds to the heat of compression of hydrogen which is generated due to the continuous or fast filling of hydrogen for a shorter time. This phenomenon is predominant when hydrogen is filled in the Type IV storage tank of a fuel cell vehicle where the tank is a closed vessel made of composite materials. The compressed hydrogen expands inside the tank and generates a large amount of heat. The low thermal conductive materials used in the Type IV tank, such as polymer for liner, and carbon fiber reinforced polymer (CFRP) for the outer layer, restrict the heat flow inside the tank to the external environment (Table 5.1).

This results in the accumulation of a huge quantity of heat within the tank during a short filling time which eventually leads to higher-end temperature and low SOC. The quantity of heat collected within the tank also depends on refueling parameters such as filling rate and hydrogen supply temperature. The heat of compression increases as the filling rate increases although higher hydrogen supply temperature plays a role of catalyst for the same.

The total heat generated and transferred during fast refueling is critical to be estimated. Typically, reported studies have considered the overall heat transfer coefficient (HTC) to evaluate the heat transfer from the composite tank [3–6]. In general, HTC depends upon the pressure ramp rate and hydrogen inside the tank [4]. However, only the evaluation of HTC cannot provide the best results for total heat flow from the refueling station to the vehicle tank. Therefore, joint examination of heat flow with refueling parameters and standard heat transfer approach is required to have a better estimation.

DOI: 10.1201/9781003244318-5

TABLE 5.1

Thermal Properties of Tank Materials [1,2]

Materials	Density (kg/m³)	Thermal Conductivity (W/m-K)	Specific Heat (kJ/kg-K)
High density polyethylene (HDPE)	955	0.41	2.1
Carbon fiber reinforced polymer (CFRP)	1700	0.9	1400

5.2 HEAT TRANSFER IN TYPE IV TANK

The Type IV composite tank has two layers of different low-conductive materials. The layer which comes in direct contact with compressed hydrogen is the inner liner made of high-density polymer(s). Figure 5.1 shows a one-dimensional model of a Type IV tank with different modes of heat transport from hydrogen to the outer surface of the tank.

Heat transfer to the inner layer from compressed hydrogen inside the tank takes place by convention given in Equation 5.1.

$$Q = hA(\Delta T) \tag{5.1}$$

Where Q is heat transfer rate, h is convection heat transfer coefficient, A is exposed area and ΔT is temperature difference between the tank inside and the inner surface.

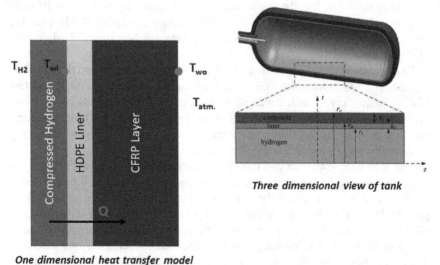

Three dimensional view of tank

One dimensional heat transfer model

FIGURE 5.1   Heat flow through the layers of Type IV compressed hydrogen storage tank.

The heat is conducted between liner and carbon fiber reinforced layer given by Equation 5.2

$$Q = -kA\left(\frac{\Delta T}{\Delta x}\right) \tag{5.2}$$

Where Q heat transfer rate, k is conductivity of materials, A is exposed area and $\frac{\Delta T}{\Delta x}$ is temperature gradient between two layers. Heat discharge to the atmosphere from the composite layer by convection is given by Equation 5.1.

The mass of hydrogen entered inside the tank in a compressible manner. Due to the compressible filling of hydrogen the velocity of flow, gravitational forces and inertia cannot be ignored as these forces act on the gas. For this, a dimensional number such as Reynolds number, Froud number and Mach number need to be considered for the examination of the flow behavior of compressed gas (Equations 5.3–5.5)

$$\text{Reynolds number: } Re = \frac{\rho dv}{\mu} \tag{5.3}$$

$$\text{Froud number: } Fr = \frac{u}{\sqrt{2gr}} \tag{5.4}$$

$$\text{Mach number: } Ma = \frac{u}{C_s \rho_i A} \tag{5.5}$$

Where, u is velocity, C speed of sound, ρ_i density, g gravity and μ viscosity.

The transfer of heat from compressed hydrogen to the liner happens only by convection but it cannot be natural convection as the compressible flow of gas within the tank is involved when the refueling was under process. Therefore, the mode of heat transfer can be a combination of natural and forced convection defined in terms of Nusselt's number and Rayleigh number given by Equations 5.6–5.9.

$$Nu_{\text{Forced}} = C_w Re^{1/2} \tag{5.6}$$

$$Nu_{Free} = C_w Ra^{1/3} \tag{5.7}$$

$$Ra = \frac{g\beta(T - T_{wi})\rho^2 D^3 C_p}{\mu^2 k} \tag{5.8}$$

$$Nu = \frac{hd}{k} = \left(h_{Forced}^4 + h_{Free}^4\right)1/4 \tag{5.9}$$

Where Nu is Nusselt number, Re Reynolds number, Ra Rayleigh number, Cp specific heat at constant pressure and β coefficient for flow.

5.3   HEAT CAPACITY MODEL (MC METHOD)

The conventional approach to examine heat flow from the inside of the tank to the environment is not sufficient to predict the end temperature of refueling. For this, participation of refueling station and vehicle tank parameters is essential. The heat capacity model (or MC method) evaluates the heat including the end temperature and SoC of the storage system. The method developed by Honda R&D Inc., can be applied to both communication and non-communication filling stations and is known as the MC method. Where MC refers to the mathematical construct or total heat capacity of the system which comprises combined specific heat and mass capacity of the tank [7,8]. If actual thermodynamics conditions at the station and the vehicle tank are considered, exact tank filling outcomes can be obtained.

Figure 5.2 represents heat flow from the refueling station to the vehicle tank. The total heat conveyed to the vehicle tank depends on these two factors – first, temperature and pressure of refueling station reservoir while the second is the change in temperature during flow through different units of the station to the vehicle tank.

Figure 5.3 shows the total energy transferred to the tank from the refueling station and the heat capacity of tank walls. The fluid domain of the tank has been considered as control volume, and walls are considered as a characteristic volume for energy and heat transfer calculations. The integration of station and vehicle tank parameters for assessment of heat transferred and end temperature of refueling is a major advantage of implementing the heat capacity model.

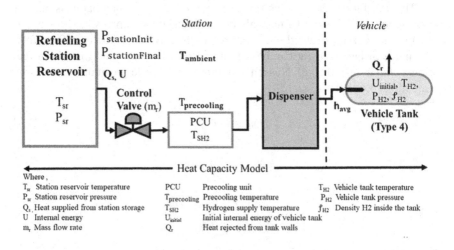

FIGURE 5.2   Heat flow schematic of refueling station to vehicle tank.

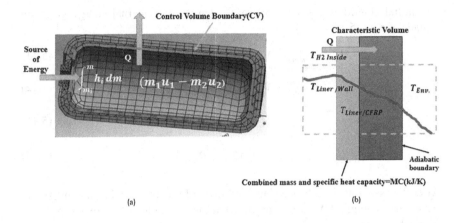

FIGURE 5.3 Schematic of energy and heat flow during refueling (a) energy transferred in control volume (b) temperature distribution on a section of the wall.

The requisite parameters for the development of the heat capacity model and assessment of final temperature from the station and the vehicle are tabulated in Table 5.2. Based on the station and vehicle refueling parameter, total heat transfer, internal energy and quantity of hydrogen transferred from the refueling station to the vehicle tank can be estimated.

The HCM needs the information from the vehicle tank as it is based on communication type refueling stations. For this, the refueling station takes initial conditions of the temperature and the pressure from the vehicle tank mainly. The initial temperature ($T_{initial}$) of the storage system is given by Equation 5.10. The initial temperature ($T_{initial}$) is the addition of ambient ($T_{ambient}$) and hot soak temperature (ΔT_{hot}), shown by Equation (5.10).

$$T_{initial} = T_{ambient} + \Delta T_{hot} \qquad (5.10)$$

TABLE 5.2
Station and Vehicle Tank Parameters [9]

Station	Vehicle
Ambient temperature	Hot soak temperature
Storage pressure, temperature	Tank initial pressure, temperature and density
Enthalpy of H_2 supplied at storage pressure and temperature	Volume of the tank
Station type (A, B, C, D)	Initial internal energy

The initial condition also includes mass, density and internal energy given by Equations 5.11–5.14.

$$m_{initial} = V_{vt} \times \rho_{initial}\left(T_{initial}, P_{Sinitial}\right) \tag{5.11}$$

$$m_{CV} = V_{vt} \times \rho_{target} \tag{5.12}$$

$$m_{add} = m_{cv} - m_{initial} \tag{5.13}$$

$$u_{initial} = u_{initial}\left(T_{initial}, P_{initial}\right) \tag{5.14}$$

After the assessment of the starting conditions of the tank, total heat conveyed to the control volume (CV) was determined at the supply temperature and pressure of the hydrogen stream. The average enthalpy conveyed to the vehicle tank is determined via the Runge–Kutta approximation method as depicted by Equation (5.15).

$$
\begin{aligned}
\underline{h} = \frac{1}{4} &\left[\frac{h\left(T_{pc}, \; P_{SI}\right) + h(T_{pc}, \left(P_{SI} + \frac{P_{SF} - P_{SI}}{4}\right)}{2} \right] \\
&+ \left[\frac{h(T_{pc}, \; \left(P_{SI} + \frac{P_{SF} - P_{SI}}{4}\right) + h(T_{pc}, \left(P_{SI} + 2\frac{P_{SF} - P_{SI}}{4}\right)}{2} \right] \\
&+ \left[\frac{h(T_{pc}, \; \left(P_{SI} + 2\frac{P_{SF} - P_{SI}}{4}\right) + h(T_{pc}, \left(P_{SI} + 3\frac{P_{SF} - P_{SI}}{4}\right)}{2} \right] \\
&+ \left[\frac{h(T_{pc}, \; \left(P_{SI} + 3\frac{P_{SF} - P_{SI}}{4}\right) + h(T_{pc}, \left(P_{SI} + 4\frac{P_{SF} - P_{SI}}{4}\right)}{2} \right]
\end{aligned}
\tag{5.15}
$$

Where Tpc, P_{SI} and P_{SF} are expected precooling temperature, initial pressure at the station and final pressure at the station, respectively.

Considering the average enthalpy supplied ($\underline{h}$) to tank, adiabatic internal energy ($U_{adiabatic}$) and adiabatic temperature ($T_{adiabatic}$) can be calculated for the adiabatic filling condition where heat transferred from the system is zero. The combined mass and specific heat capacity of the characteristic volume are calculated using Equation (5.16) which represents the total heat absorbed by the wall.

$$MC = C + A\frac{U_{adaibatic}}{U_{initial}} + g\left(1 - e^{-k\Delta t}\right)^{j} \tag{5.16}$$

Where the empirical coefficients (C, A, g, e, k, j) depicted in Equation (5.16) for estimation of MC are obtained from the confirmation test performed on Type IV tank where the filling time is more than 3 minutes [31]. The adiabatic internal energy of control volume defined by Equation (5.17) is

$$U_{adiabatic} = \frac{m_{initial}U_{initial} + m_{add}h_{avg}}{m_{cv}} \tag{5.17}$$

Ultimately, the end temperature (T_{end}) of the gas is obtained by utilizing tank parameters, initial temperature, specific heat (C_v), adiabatic temperature ($T_{adiabatic}$), mass (m_{cv}) and heat capacity (MC) for filling time of 200 seconds are correlated by Equation (5.18).

$$T_{end} = \frac{m_{cv}C_vT_{adiabatic} + MCT_{initial}}{\left(MC + m_{cv}C_v\right)} \tag{5.18}$$

5.4   CASE STUDY: IMPLEMENTATION OF HCM IN ADIABATIC SIMULATION OF 29 L TYPE IV TANK [10]

The refueling simulation discussed in Chapter 4 is an adiabatic simulation of the refueling process in a 29 L Type IV tank. The temperature rise was high and the state of charge obtained at different filling conditions was lesser than desired. This is due to the neglect of heat transfer in the entire process of refueling. But the transfer of heat in a compressible flow at a high temperature cannot be ignored. For this, HCM is applied to adiabatic refueling simulation to examine the end temperature and state of charge attained.

The heat capacity model was applied to evaluate the heat delivered, internal energy and heat capacity followed by end temperature and the SoC. The total heat transferred to the vehicle tank from the refueling station depends upon the pressure, temperature and mass of compressed hydrogen delivered from the station reservoir. The supplied pressure of hydrogen was similar to the storage pressure which was 87.5 MPa for operating pressure of 70 MPa and the supply temperature varies from −40°C to 15°C for different station types.

The total heat transferred also depends on the initial state of the vehicle tank such as initial pressure and temperature. In this study, the initial pressure of the tank is 2 MPa, and the initial temperature of the tank is 20°C. It was presumed that the vehicle is in a hot soak condition where the environment possesses a higher temperature than ambient conditions. The hot soak margin of safety to overheat has taken to be +7.5°C [8]. Therefore, the total heat delivered from the refueling station to the vehicle tank is designated as average enthalpy obtained from Equation 5.15.

Table 5.3 presents the average enthalpy delivered to the vehicle tank at different supply temperatures evaluated by using the Runge–Kutta approximation.

The total energy content or the adiabatic internal energy of control volume is evaluated using Equation 5.17 which depends on initial internal energy, mass of

TABLE 5.3
Average Enthalpy Delivered to the Vehicle Tank at Different Supply Temperatures

Hydrogen Supply Temperature (°C)	Average Enthalpy Delivered (kJ/kg)
−40	6870.87
−20	7471.84
0	8070.62
15	8518.31

TABLE 5.4
Available Internal Energy in Control Volume of 29 L Tank

Hydrogen Supply Temperature (°C)	Adiabatic Internal Energy (kJ/kg)
−40	7019.31
−20	7621.35
0	8221.21
15	8669.70

hydrogen added and mass average enthalpy delivered to the vehicle tank. Table 5.3 presents adiabatic internal energy of control volume of tank for different hydrogen supply temperatures.

Based on adiabatic internal energy and initial internal energy of the vehicle tank heat capacity (MC) of characteristic volume has been calculated using Equation 5.16, as given in Table 5.5 at different supply temperatures of hydrogen. The heat transferred from compressed hydrogen to the tank walls is less for low supply temperature and increases with higher supply temperature. Therefore, MC has a smaller value at precooled supply temperature and a higher value at 0 and 15°C.

Figure 5.4 presents the temperature obtained at the end of refueling after the implementation of the heat capacity model. For supply temperature of 15°C, the

TABLE 5.5
Heat Capacity (MC) of Characteristic Volume of 29 L Tank

Hydrogen Supply Temperature (°C)	Heat Capacity (kJ/K)
−40	12.34
−20	15.67
0	18.39
15	22.23

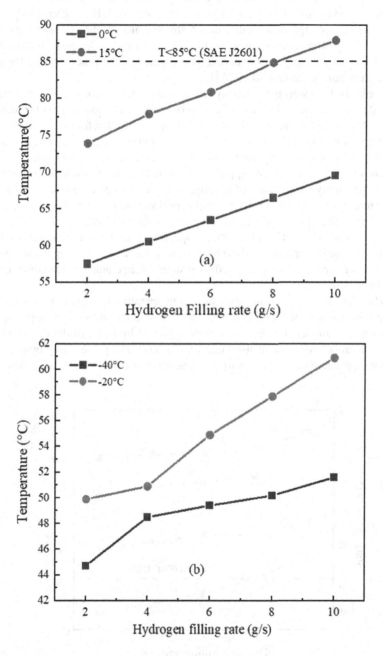

FIGURE 5.4 End temperature using heat capacity model for 29 L tank at different filling rates and supply temperatures of (a) 0°C and 15°C (b) 20°C and −40°C.

end temperature remains lower for low filling rates but still lies above the limiting value defined by SAE J2601 for 10 g/s as in Figure 5.4(a). For higher filling rates, compressed gas requires a smaller time frame to transport the produced heat during refueling. The analysis, done by Liu and co-workers., suggests that the heat transfer requires a larger time period as compared to the time required for heat generation during the fast filling [11].

Similarly, by lowering the supply temperature, a lesser amount of heat is transported to the vehicle tank that restricts the rise of the end temperature of refueling. For supply temperature of 0°C, the end temperature of refueling decreases by 16–18°C at different filling rates as shown in Figure 5.4(a) and also restricts the end temperature to recommend a limit of 85°C even at faster filling rates.

The effect of lowering of supply temperature and heat transfer is more visible when the supply temperature of hydrogen decreases to precool level as depicted in Figure 5.4(b), which causes a significant decrease in the end temperature of refueling at all filling rates. It lies in the range of 49–61°C and 43–52°C for supply temperatures of −20°C and −40°C, respectively, as shown in Figure 5.4(b). Therefore, the incorporation of the heat capacity model in refueling decreases the end temperature to 1.3–2.4°C at different supply temperatures when compared to adiabatic simulation.

The SOC is expressed in terms of end temperature and pressure. Figure 5.5 shows the effect of filling rate and supply temperature on SOC, by applying the heat capacity model. The SOC has been improved by virtue of the decrement in the end temperature, but at the same time, it does not possess any influence at upper supply temperature and still lies beneath 90% even at low filling rates.

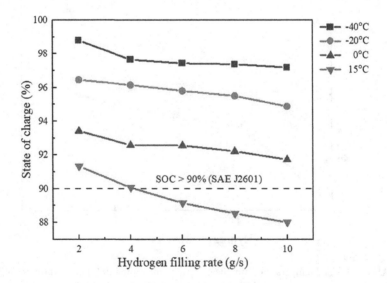

FIGURE 5.5 State of charge of 29 L tank achieved using heat capacity model at various supply temperature and hydrogen filling rates.

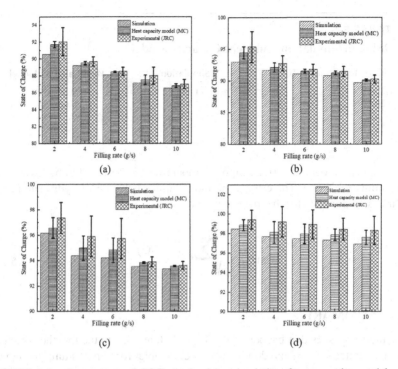

FIGURE 5.6 Comparison of SOC obtained by simulation, heat capacity model and experimental (JRC) at different hydrogen supply temperatures of (a) 15°C (b) 0°C (c) −20°C (d) −40°C.

Figure 5.6 shows an appraisal of the SoC obtained after implementation of heat capacity model in adiabatic refueling simulation with experimental results of GasTeF at JRC [12]. The state of charge has shown improvement and inches toward the experimental results. Figure 5.6 also presents the corresponding error before and after the implementation of the heat capacity model as given by Equations 5.19 and 5.20.

$$e = \left(\rho_{Experimental} - \rho_{Simulation}\right)/\rho_{experimental} \tag{5.19}$$

$$e = \left(\rho_{Experimental} - \rho_{HCM}\right)/\rho_{Experimental} \tag{5.20}$$

The standard error between experimental and heat capacity model results has been reduced to less than 1% for most of the supply temperature. The initial and boundary conditions put in the calculation correspond to such trivial deviations that occurred in results.

For more logical assessments of the results, root means square error (RMSE) is obtained by using Equation 5.21 and is displayed in Table 5.6. The RMSE of the

TABLE 5.6
RMSE in Simulation and Heat Capacity Model

Supply Temperature (°C)	Simulation	Heat Capacity Model
−40	1.31530	0.805179
−20	1.11964	0.682174
0	1.27325	0.539914
15	0.85713	0.298061

heat capacity model at various supply temperatures is observed to be lesser than the simulation's one. The critical thermodynamics behavior of hydrogen during filling is responsible for the low state of charge.

$$RMSE = \sqrt{\frac{\sum_{i=1}^{n}(x_1 - x_2)^2}{n}} \tag{5.21}$$

5.5 SUMMARY

The thermal issues that happen in the Type IV tank, due to the low thermal conductive materials, are predominantly accountable for temperature increment inside the tank. The real behavior of hydrogen and heat of compression at high pressure and temperature result in high-end temperature and lower SOC. The heat flow investigation requires both refueling station and vehicle tank parameters to examine the temperature inside the tank, exactly. For this, a heat capacity model has been developed by integrating refueling station and vehicle tank parameters. The results of the case study discussed show a good improvement in the state of charge with the implementation of the heat capacity model(s) in the refueling process.

REFERENCES

1. He, C.Y., R. Yu, H. Sun, Z. Chen, Lightweight multilayer composite structure for hydrogen storage tank. *International Journal of Hydrogen Energy*, 41, 15812–15816, 2016.
2. Roh, H.S., T.Q. Hua, R.K. Ahluwalia, Optimization of carbon fiber usage in Type 4 hydrogen storage tanks for fuel cell automobiles. *International Journal of Hydrogen Energy*, 38(29), 12795–12802, 2013.
3. Zhao, L., et al., Thermodynamic analysis of the emptying process of compressed hydrogen tanks. *International Journal of Hydrogen Energy*, 44(7), 3993–4005, 2019.
4. Wang, L., et al., Heat transfer analysis for fast filling of on-board hydrogen tank. *Energy Procedia*, 158, 1910–1916, 2019.
5. Molkov, V., M. Dadashzadeh, D. Makarov, Physical model of onboard hydrogen storage tank thermal behaviour during fuelling. *International Journal of Hydrogen Energy*, 44(8), 4374–4384, 2019.

6. Kim, S.C.L., S.H. Yoon, K. Bong, Thermal characteristics during hydrogen fueling process of type IV cylinder. *International Journal of Hydrogen Energy*, 35(13), 6830–6835, 2010.

7. Harty, R., S. Mathison, N. Gupta. Improving hydrogen tank refueling performance through the use of an advanced fueling algorithm – the MC method. in *National Hydrogen Association Conference*, Long Beach, CA: Honda R&D Americas, Inc., 2010.

8. Mathison, S., K. Handa, T. McGuire, T. Brown, T. Goldstein, M. Johnston, Field validation of the MC default fill hydrogen fueling protocol. *SAE International Journal of Alternative Powertrains*, 1(4), 131–144, 2015.

9. Reddi, K., A. Elgowainy, N. Rustagi, E. Gupta, Impact of hydrogen SAE J2601 fueling methods on fueling time of light-duty fuel cell electric vehicles. *International Journal of Hydrogen Energy*, 42(26), 16675–16685, 2017.

10. Sapre, S., K. Pareek, R. Rohan, P.K. Singh, H_2 refueling assessment of composite storage tank for fuel cell vehicle. *International Journal of Hydrogen Energy*, 44(42), 23699–23707, 2019.

11. Liu, Y.-L., et al., Experimental studies on temperature rise within a hydrogen cylinder during refueling. *International Journal of Hydrogen Energy*, 35(7), 2627–2632, 2010.

12. Acosta, B., P. Moretto, N. de Miguel, R. Ortiz, F. Harskamp, C. Bonato, JRC reference data from experiments of on-board hydrogen tanks fast filling. *International Journal of Hydrogen Energy*, 39(35), 20531–20537, 2014.

6 Structural Analysis of Type IV Tanks

6.1 INTRODUCTION

The Type IV tank at pressures of 35 MPa and 70 MPa is an encouraging preference, as it possesses high "strength/stiffness-to-weight ratio," and outstanding protection ability against fatigue and corrosion [1,2]. The synergic effect of high density polyethylene (HDPE) liner and carbon fiber reinforced polymer (CFRP) composite brings value addition to the tanks used for fuel cell vehicle applications [3]. However, different refueling conditions cause deviations to pressure and temperature for a short period of time. This cyclic load weakens the composite structure by degrading either polymer liner or carbon fiber/resin bonds in laminates, leading to the damage of the whole structure.

Significant research has been performed to get an understanding of mechanical and thermal responses of the composite tanks by applying theoretical models as well as conducting experiments in the last several years [4–10]. A number of new researches and analyses have been done for materials characteristics [11], burst pressure [8,12,13], stress and strain response [14,15], strength analysis [16], failure at different pressure [6,17], etc.

Type IV tank for compressed hydrogen storage operates at a nominal working pressure of 70 MPa for fuel cell vehicles. Currently, all manufacturers like Quantum, Hexagon and Lincoln fabricate the Type IV tanks with the factor of safety ranging from 1.25–2.25 times of nominal working pressure. However, frequent charging and discharging cycles sometimes generate severe operating conditions. The fast charging of the tank generates higher pressure ramp rates. The pressure rates along with the heat of compression are responsible for the increase in temperature inside the tank. The internal heat generated due to the compression depends on the filling rates and hydrogen supply temperature.

Apart from this, undefined loads produced by the motion on compressed hydrogen inside the tank such as thermal stratification, hot spots and irregular pressure rise are difficult to be examined. These undesirable phenomena affect the mechanical and thermal behaviors of materials which cause failure of the tank materials. The failure of the tank originates when a decrease of strength in tension and compression occurs due to high pressure and temperature, which corresponds for the structural stability of the tank, at abnormal operating conditions.

DOI: 10.1201/9781003244318-6

6.2 ANALYSIS OF TYPE IV TANK

For analysis, the Type IV tank is subjected to severe internal loading conditions which eventually affect the structure of the tank. These internal loads may be frequent or cyclic, depending upon the refueling conditions and rate of pressure and temperature rise. Generally, cyclic loading is more predominant when the tank is fast-charged in a shorter time. Therefore, it is pertinent to analyze the effect of cyclic loading on a composite structure which assists to understand the behavior of the tank.

The internal compressive loads produce stress, strain and deformation on the different regions of the tank. This distribution is measured in circumferential, longitudinal and radial directions as shown in Figure 6.1. The distribution of stress, strain and deformation are mainly observed at three sections of the tank i.e. cylindrical, dome and junction of dome and cylinder.

The region of maximum stress or deformation originates the chances of failure of the tank in the respective locations. The appearance of the crack(s) that lead to breakage of ply or lamina is called as the failure of the tank. The tank offers resistance to internal loadings depending upon the strength of the materials. However, at high pressure and temperature conditions, the chances of failure are more. The breakage of fiber, delamination and fiber-matrix debonding are some possible reasons that cause failure in Type IV composite tanks. In general, composite structure may fail if the strength ratio is less than 1 (one) which is defined as the ratio of maximum load to the applied load.

For the assessment of failure, theories of failures can be used to predict the durability of a structure. This can be done by applying various theories of failure defined in solid mechanics. Failure theories like "Maximum stress/Principal stress" and "Maximum strain/Principal strain" are used to predict the conditions under which a solid material fails when subjected to the loadings.

In general, the maximum stress and strain theories correspond for tensile strength during the yielding. The failure occurs when maximum stress along

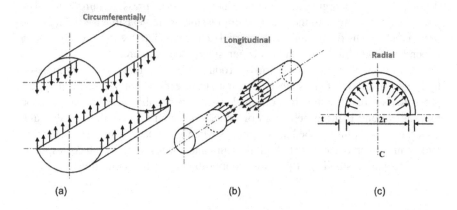

FIGURE 6.1   (a) Circumferential (b) Longitudinal (c) Radial in a composite tank.

one axis attains ultimate or yield strength The failure criteria are given by Equation 6.1.

$$X_C < \sigma_1 < X_T \quad X_C < \sigma_2 < X_T \quad X_C < \sigma_3 < X_T \tag{6.1}$$

Where X_C, X_T represent the compressive and tensile strength, respectively and σ_1, σ_2 and σ_3 are ultimate stress in three directions. It provides the maximum stress (compressive or tensile) in a particular direction.

In the same way, yielding happens when the maximum principal strain crosses the strain at the tensile yield point in either simple tension or compression. The failure criteria are depicted by Equation 6.2.

$$\epsilon_C < \epsilon_1 < \epsilon_T \quad \epsilon_C < \epsilon_2 < \epsilon_T \quad \epsilon_C < \epsilon_3 < \epsilon_T \tag{6.2}$$

Where, ϵ_C, ϵ_T represent the compressive and tensile strain, respectively and ϵ_1, ϵ_2 and ϵ_3 represent ultimate strain in three directions. It provides the maximum strain (compressive or tensile) in a particular direction.

The equivalent stress and strain theory used to predict the stress and strain at critical loading conditions usually considers tensile yield stress.

$$\epsilon_1 = \frac{1}{E}\left(\sigma_1 - V\sigma_2\right) \tag{6.3}$$

$$\epsilon_2 = \frac{1}{E}\left(\sigma_2 - V\sigma_1\right) \tag{6.4}$$

Where σ_1 and σ_2 are stresses in two different directions, ϵ_1 and ϵ_2 are strains in the corresponding direction.

For the composite tank, failure theories can be applied to the individual plies of the tank. The mechanical and thermal stresses produced on each ply are evaluated. Based on the maximum stress and strain the failed ply is identified. The failure pressure of the tank, also called burst pressure, is evaluated based on the first ply failure.

The burst of the tank can be categorized based on its origin and severity. When failure occurs inside the cylindrical portion and the metallic bosses move inside, the tank is called a safe burst, as shown in Figure 6.2 [18]. Fiber breaking occurring on circumferential plies is predominantly responsible for such bursts. The modes of damage e.g. matrix-cracking & delamination contribute less significantly for these types of bursts.

When metallic boss ejection occurs because of failure happened in dome, it is called as unsafe as shown in Figure 6.2 [18]. Though fiber breakage of helical plies is mainly responsible for such failures, matrix cracking can be accountable to decide the mode of burst from safe to unsafe because of overloading of fibers.

(a)   (b)

FIGURE 6.2   (a) Safe and (b) unsafe burst of Type IV tanks [18].

6.3 FAILURE ANALYSIS OF TYPE IV TANK

Type IV tank, when subjected to cyclic loading, the failure can be observed by the appearance of stress, strain and deformation. Stress, strain and deformation are generally observed in three regions of the tank mainly cylindrical, dome and junction of the tank. The effect of operating conditions is first influenced by a liner that comes in direct contact with compressed hydrogen. Then, the internal loads are transferred to the composite layers. Therefore, chances of failure are more in liner and immediate layers of composite materials. In this section, failure is analyzed based on the refueling results of Chapter 4 for a 29 L Type IV tank at different filling conditions. It includes the effect of pressure and temperature attained at the end of refueling at various supply temperatures and filling rates.

6.3.1 Effect of Cyclic Loading on LINER

The high-pressure compressed hydrogen directly comes in contact with HDPE liner. The stress and strain produced on the surface of the liner have been evaluated using different theories of failure applied on the finite element model of the tank as shown in Figure 6.3. The evaluation of stress and strain was performed on nominal working pressure to maximum pressure attained during the refueling at different filling rates. The appearance of maximum stress on the surface of the liner decides the probability of failure of the liner. The stress produced on the liner at different loads should be less than the ultimate strength of HDPE to avoid the failure of the liner.

Figure 6.4 shows the distribution of stress on the liner surface at a nominal working pressure of 70 MPa and 15°C. The maximum and minimum stress appeared on the cylindrical and dome portion of the tank, respectively. This confirms that stress can cause damage in the cylindrical part if stress exceeds beyond the ultimate strength of HDPE which is 25 MPa. Nevertheless, as published in Reference [19], it can be called "safe burst." The maximum stress calculated using principal and equivalent stress failure criteria is converged to 20.28 MPa and 17.04 MPa, respectively, which corresponds to lesser values than yield stress values of HDPE liner at nominal conditions of pressure and temperature.

Figure 6.5 (a and b) show strain produced at 70 MPa and 15°C using principal elastic strain and equivalent strain failure criteria, respectively. The appearance of stress confirms that the maximum strain would be observed in the cylindrical portion only which displays the probability of deformation at this portion, and can be

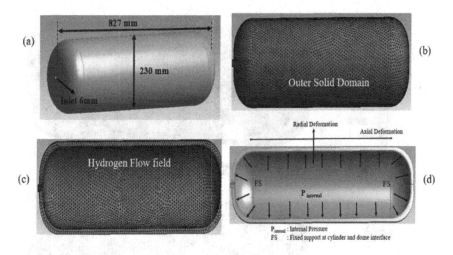

FIGURE 6.3 (a) 3-D model of Type IV tank (b) and (c) Computational model for inner and outer region of the tank (d) Internal loads in tank and deformation.

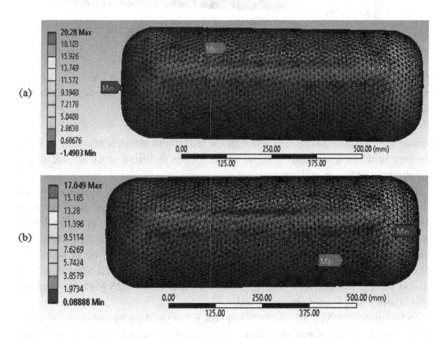

FIGURE 6.4 Appearance of stress on the liner of Type IV tank at a nominal working pressure of 70 MPa and temperature 15°C using (a) maximum principal stress theory (b) equivalent (Von Mises) stress theory.

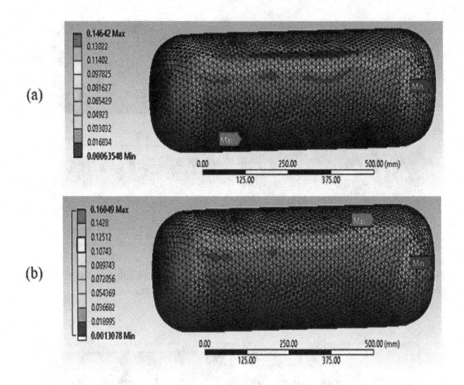

FIGURE 6.5 Appearance of strain on the liner of Type IV tank at a nominal working pressure of 70 MPa and temperature 15°C using (a) maximum principal elastic strain theory (b) equivalent elastic strain (Von Mises) theory.

termed as "radial deformation." Though the maximum stress at nominal working pressure remains lesser than the final value, it rises as filling proceeds because of variations in pressure and temperature.

Figure 6.6(a) and (b) shows the stresses and strains obtained at different pressure found when the refueling takes place at a filling rate of 2–10 g/s. Figure 6.6 also shows the stress and strain developed at maximum allowable pressure demarcated by SAE J2601 protocol, which is 125% of nominal operating pressure. The maximum stress of 21.95 MPa and 25.45 MPa were found to be at maximum allowable pressure of 87.5 MPa using maximum stress theory and equivalent (Von Mises) theory, respectively. The stress crosses the limit of 25 MPa by principle stress theory at allowable pressure only.

In Figure 6.6(a), Von Mises criteria envisage the yielding by considering tensile and compressive stress and, infer that the stress lies within the range. The polymers generally have higher compressive stress than tensile. Therefore, the maximum stress obtained by Von Mises criteria was lower than the maximum stress criteria. However, maximum stress theory gives more real value of stress than equivalent stress theory which only gives a theoretical measure of stress [20]. Figure 6.6(b)

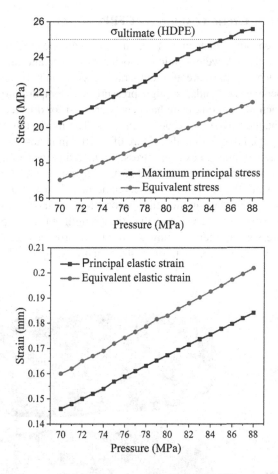

FIGURE 6.6 (a) Stress and (b) strain produced on the liner at different refueling pressures.

represents the strain at different pressures using two different criteria for failure. The linear variation of strain is similar to the stress which concluded that the deformation will appear on the liner where the stress is maximum. However, the high tensile and compressive strength of HDPE restrict the deformation of liner [21].

In addition, the temperature generated during the refueling is also responsible for deformation, but due to the high coefficient of thermal expansion and high stiffness, it affects the deformation of the liner in a very limited manner. Another reason for the limited deformation of HDPE was the support provided by the strong binding of carbon fiber on its surface. The simulation performed in this study suggests a small deformation in HDPE in the cylindrical part of the tank. The reported experimental studies on HDPE liner also suggested the influence of temperature on HDPE that appears at temperatures above 80–90°C [22,23]. Therefore, the probability of failure of the composite layer is higher than the failure of the HDPE liner.

6.3.2 EFFECT OF CYCLIC LOADING ON CFRP

The reinforcement strength decides the failure of the composite structure. Though CFRP has high strength/weight ratio, corrosion resistance and fatigue resistance they get influenced by pressure and temperature, which may be attributed to the ruining of fiber-matrix bonds, at high pressure and temperature. For indepth explanation, each ply of composite has been inspected for different pressure and temperature obtained from the refueling process. The maximum principal stress criteria were applied to examine the failure of CFRP laminates.

Figure 6.7(a), (b), and (c) exhibits stress, strain and deformation on ply 1 at a nominal working pressure of 70 MPa and 15°C using maximum stress theory. The maximum stress of 106.46 MPa was observed at the cylindrical part, which leads to maximum strain and deformation in this portion of the tank as in Figure 6.6(b) and (c). The maximum stress was less than the strength of the composite at nominal working pressure and temperature. As refueling proceeds pressure increases the filling rates leading to the increase in stress on the composite layer.

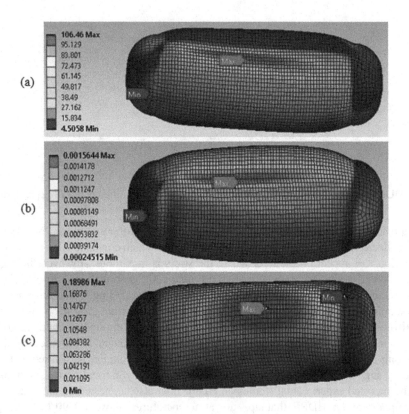

FIGURE 6.7   (a) Stress (b) strain and (c) total deformation on carbon fiber/epoxy ply 1 of the tank at nominal working pressure (70 MPa) and temperature 15°C.

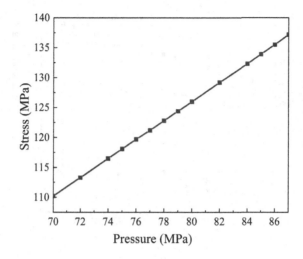

FIGURE 6.8   Stress on ply 1 of the tank at different pressure and temperature 15°C.

Figure 6.8 illustrates the stress produced at different refueling pressures attained at filling rates of 2–10 g/s. It displays a constant increment in stress along with the pressure of the tank. The maximum stress of 137.20 MPa was noticed at a pressure of 87 MPa which is the maximum allowable pressure demarcated by SAE J2601 for safe refueling of the tank. It was also the maximum stress which can be afforded by the fiber of ply 1. This value was compared with an experimental burst pressure for T700s which is 148.30 MPa as fiber is the only load-carrying element in composite [18,24].

Figure 6.9 shows the influence of refueling pressure on various plies of the stack. The increment in pressure corresponds to increment in stress, as the refueling increases; the distribution of stress on different plies resembles to ply 1, as shown in Figure 6.8. The deformation seems safe as it occurs on the cylindrical surface of the tank. Figure 6.9 also infers that ply 1 would be the first which undergoes failure, happening due to increment in either pressure or temperature.

The effect of temperature on CFRP composite cannot be overlooked. During the refueling of compressed hydrogen, temperature increases exponentially with pressure and the filling rates. The temperature has a significant consequence on the fiber-matrix interface and adhesion of CF on the liner, which leads to the failure of ply. The fiber-matrix bonding may deteriorate in extreme cold and hot conditions generated within the pressure vessel during refueling.

The effect of temperature is experienced in ply 1 exclusively, as it has shown the great possibility of failure with variation in pressure range. For such purpose, ply 1 was investigated for the temperature obtained at different filling rates and supply temperatures.

Figure 6.10 displays the thermal stress at the end temperature attained during the refueling at 2–10 g/s for supply temperatures of –40–15°C. The end temperature possesses robust relation with filling rates and supply temperature. Figure

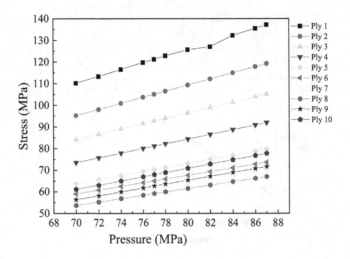

FIGURE 6.9   Maximum stress in plies of the composite structure at different pressure.

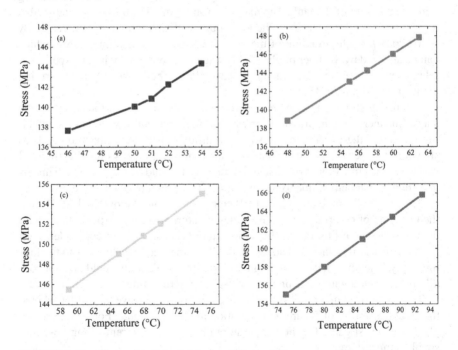

FIGURE 6.10   Thermal stresses in ply 1 at different filling rates and supply temperatures of (a) –40°C (b) –20°C (c) 0°C and (d) 15°C.

6.10(a)–(d) shows the linear increase in stress as filling rates increases with supply temperature. Hence, maximum stress was observed at a supply temperature of 15°C and a filling rate of 10 g/s.

6.4 BURST PRESSURE

The mechanical and thermal analysis of HDPE liner and CFRP laminate in previous sections mainly focus on the distribution of stress and location of maximum stress on the tank surface. The results confirm that the most probable location of failure of the tank is the cylindrical portion of the tank. The maximum stress and deformation at the respective pressure, temperature and filling rates during the fast-filling process are tabulated in Table 6.1. Since SAE J2601 suggests the operating range of pressure and temperature as 87.5 MPa and 85°C for safe refueling, the pressure 165.81 MPa was considered as burst pressure for the composite tank.

Table 6.2 shows a comparative analysis of burst pressure of the Type IV tank used in this study with the tank having some similar design parameters used in previous experimental and simulation analysis.

The burst pressure obtained by this study shows a good agreement with a small variation of 0.68–5.25% with experimental and 2.29–4.71% with simulation results.

TABLE 6.1
Maximum Stresses and Deformation at Various Filling Conditions

Filling Rate (g/s)	Maximum Pressure (MPa)	Maximum Temperature (°C)	Maximum Stress (MPa)	Deformation (mm)
2	77	75	155.04	2.1917
4	77.6	80	158.03	2.2166
6	78.5	85	161.02	2.2203
8	79.3	89	163.42	3.3919
10	79.8	93	165.81	4.4205

TABLE 6.2
Comparison of Burst Pressure

Burst Pressure (MPa)	Remark	Difference (5)	Reference
174.00	Experimental	4.70	[25]
166.95	Experimental	0.68	[26]
172.90	Experimental	4.10	[27]
175.00	Experimental	5.25	[18]
162.00	Simulation	2.29	
160.00	Simulation	3.50	[28]
158.00	Simulation	4.71	[29]
165.81	**Simulation**		**Present study**

The experimental results show higher burst pressure compared to the simulation studies. It may be due to the different operating conditions, material properties, type of carbon fiber used in laminate and type of polymer used in the liner. Similarly, modeling assumption, approach, failure consideration at tank, laminate and ply level make the difference in burst pressure of the tank obtained by simulation.

6.5 SUMMARY

The Type IV tank is subjected to severe internal loading conditions when used in onboard applications. The internal load is the result of the fast charging of Type IV at different filling conditions. The structure of the tank is subjected to intense stresses leading to the deformation of the tank in a radial and axial direction. However, the deformation is either safe or unsafe depending on the movement of the boss. The failure of the tank is defined in terms of burst pressure based on the maximum mechanical and thermal stress tolerated by the tank. The structure is mechanically and thermally stable till the mechanical and thermal properties of materials cannot be affected. The Type IV tank has shown good structural stability at high pressure and temperature of refueling.

REFERENCES

1. Abe, J.O., et al., Hydrogen energy, economy and storage: Review and recommendation. *International Journal of Hydrogen Energy*, 44(29), 15072–15086, 2019.
2. Daghia, F., et al., A hierarchy of models for the design of composite pressure vessels. *Composite Structures*, 235, 111809, 2020.
3. Zhang, Q., et al., Design of a 70 MPa type IV hydrogen storage vessel using accurate modeling techniques for dome thickness prediction. *Composite Structures*, 236, 111915, 2020.
4. Dadashzadeh, M., et al., Risk assessment methodology for onboard hydrogen storage. *International Journal of Hydrogen Energy*, 43(12), 6462–6475, 2018.
5. Han, M.-G., S.-H. Chang, Failure analysis of a Type III hydrogen pressure vessel under impact loading induced by free fall. *Composite Structures*, 127, 288–297, 2015.
6. Wang, L., B. Wang, S. Wei, Y. Hong, C. Zheng, Prediction of long-term fatigue life of CFRP composite hydrogen storage vessel based on micromechanics of failure. *Composites Part B: Engineering*, 97, 274–281, 2016.
7. Liu, P.F., L.J. Xing, J.Y. Zheng, Failure analysis of carbon fiber/epoxy composite cylindrical laminates using explicit finite element method. *Composites Part B: Engineering*, 56, 54–61, 2014.
8. Onder, A., O. Sayman, T. Dogan, N. Tarakcioglu, Burst failure load of composite pressure vessels. *Composite Structures*, 89(1), 159–166, 2009.
9. Wang, L., et al., Micromechanics-based progressive failure analysis of carbon fiber/ epoxy composite vessel under combined internal pressure and thermomechanical loading. *Composites Part B: Engineering*, 89, 77–84, 2016.
10. Zhao, Y., et al., Numerical study on fast filling of 70 MPa type III cylinder for hydrogen vehicle. *International Journal of Hydrogen Energy*, 37(22), 17517–17522, 2012.

11. Parnas, L., N. Katırcı, Design of fiber-reinforced composite pressure vessels under various loading conditions. *Composite Structures*, 58(1), 83–95, 2002.

12. Blanc-Vannet, P., Burst pressure reduction of various thermoset composite pressure vessels after impact on the cylindrical part. *Composite Structures*, 160, 706–711, 2017.

13. Mair, G.W., et al., Monte-Carlo-analysis of minimum load cycle requirements for composite cylinders for hydrogen. *International Journal of Hydrogen Energy*, 44(17), 8833–8841, 2019.

14. Badri, T.M., H.H. Al-Kayiem, Numerical analysis of thermal and eleastic stress in thick pressure vessel for cryogenic hydrogen sstorage application. *Journal of Applied Sciences*, 11,. 1756–1762, 2011.

15. Wu, Q.G., et al., Stress and damage analyses of composite overwrapped pressure vessel. *Procedia Engineering*, 130(Supplement C), 32–40, 2015.

16. Nelson, S., et al., Verification and validation of residual stresses in composite structures. *Composite Structures*, 194, 662–673, 2018.

17. Zheng, C., S. Lei, Mechanical analysis and optimal design for carbon fiber resin composite wound hydrogen storage vessel with aluminum alloy liner. *Journal of Pressure Vessel Technology*, 131(2), 021204, 2009.

18. Ramirez, J.P.B., D. Halm, J.C. Grandidier, S. Villalonga, F. Nony, 700 bar type IV high pressure hydrogen storage vessel burst – simulation and experimental validation. *International Journal of Hydrogen Energy*, 40(38), 13183–13192, 2015.

19. Hamid, F., S. Akhbar, K.K. Halim, Mechanical and thermal properties of polyamide 6/HDPE-g- MAH/high density polyethylene. *Procedia Engineering*, 68, 418–424, 2013.

20. Talreja, R., Assessment of the fundamentals of failure theories for composite materials. *Composites Science and Technology*, 105, 190–201, 2014.

21. Neto, E.B., M. Chludzinski, P.B. Roese, J.S.O. Fonseca, S.C. Amico, C.A. Ferreira, Experimental and numerical analysis of a LLDPE/HDPE liner for a composite pressure vessel. *Polymer Testing*, 30(6), 693–700, 2011.

22. Merah, N., F. Saghir, Z. Khan, A. Bazoune, Effect of temperature on tensile properties of HDPE pipe material. *Plastics Rubber and Composites*, 35(4), 226–230, 2006.

23. Mahl, M., C. Jelich, H. Baier, On the temperature-dependent non-isosensitive mechanical behavior of polyethylene in a hydrogen pressure vessel. *Procedia Manufacturing*, 30, 475–482, 2019.

24. Parnas, L., N. Katırcı, Design of fiber-reinforced composite pressure vessels under various loading conditions. *Composite Structures*, 58(1), 83–95, 2002.

25. Blanc-Vannet, P., Burst pressure reduction of various thermoset composite pressure vessels after impact on the cylindrical part. *Composite Structures*, 160, 706–711, 2017.

26. Leh, D., et al., A progressive failure analysis of a 700-bar type IV hydrogen composite pressure vessel. *International Journal of Hydrogen Energy*, 40(38), 13206–13214, 2015.

27. Magneville, B., et al., Modeling, parameters identification and experimental validation of composite materials behavior law used in 700 bar type IV hydrogen high pressure storage vessel. *International Journal of Hydrogen Energy*, 40(38), 13193–13205, 2015

28. Gentilleau, B., F. Touchard, J.C. Grandidier, Numerical study of influence of temperature and matrix cracking on type IV hydrogen high pressure storage vessel behavior. *Composite Structures*, 111, 98–110, 2014.

29. Roh, H.S., T.Q. Hua, R.K. Ahluwalia, Optimization of carbon fiber usage in Type 4 hydrogen storage tanks for fuel cell automobiles. *International Journal of Hydrogen Energy*, 38(29), 12795–12802, 2013.

7 Conclusion and Future Directions

CONCLUSIONS

- The low-density hydrogen available at normal temperature and pressure needs to be stored for utilizing its energy. For this, hydrogen is stored using physical and materials-based storage. In physical storage, hydrogen is pressurized in different pressure vessels, or sometimes its temperature is lowered to its critical temperature or both can be done to increase its storage density. However, hydrogen can be stored using materials that can adsorb or absorb the hydrogen molecules to achieve the required storage density at a particular temperature. Both approaches result in higher gravimetric and volumetric density at different temperatures and pressure than normal.
- Materials-based storage system has a higher storage density than physical storage but it requires a higher temperature to be maintained for charging and discharging of hydrogen from the materials. In physical storage, hydrogen is stored in compressed, cryo-compressed and liquid forms. The boil-off and energy-intensive liquification of hydrogen restrict the commercial usage of liquid hydrogen having higher energy density. Compressed hydrogen with a storage density of 40 kg/m^3 at 70 MPa is a commercially acceptable and mature technology for onboard storage.
- Compressed hydrogen is stored in various metallic cylinders at a pressure of 20 MPa and nonmetallic cylinders up to the pressure of 70MPa. The Type IV composite cylinder is preferred for onboard storage which can store hydrogen more than 5 kg at a pressure of 70 MPa. The multilayer tank is made of high-density polymer and a composite layer of carbon fiber which provides the required strength against the highly compressed hydrogen. The Type IV tank has undergone various tests suggested by ISO, ASME and demonstrated promising results for cyclic resistance, burst pressure and hydrogen tightness. It is lacking in gravimetric and volumetric targets set by US DOE for 2020 and 2025. But Type IV tank is the first choice for FCV manufacturers for onboard hydrogen storage.
- For increased hydrogen storage density, refueling of compressed hydrogen plays a significant role in the Type IV tank. The inherently safer refueling and competing with conventional storage, refueling should follow refueling protocol i.e. SAE J2601 and SAE J2799. The temperature and pressure should be less than 85°C and 85 MPa, respectively with a

DOI: 10.1201/9781003244318-7

desired state of charge in the range of 90-100%. The success of the refueling process also depends on refueling station parameters such as filling rate, filling time, hydrogen dispensing temperature and monitoring and control by refueling interface. The sufficient exchange of information between station and vehicle can enhance the performance of the refueling process of fuel cell vehicles.

- The constant refueling approach of compressed hydrogen in Type IV tank suggests that the end temperature and pressure of refueling are lower when hydrogen is supplied at low temperature and filling rates but it may not attain the desired state of charge. Precooling of hydrogen would be required for restricting the end temperature during fast filling. Regulating the filling rate or controlled filling of hydrogen increases the state of charge by lowering the end temperature and also limits the energy consumption required for precooling.

- Fast refueling of compressed hydrogen in Type IV tanks generates a large amount of heat due to the real behavior of hydrogen and compression process. The presence of low-conductive materials in the inner and out layers causes the accumulation of heat inside the tank leading to the increase in process temperature. The estimation of heat inside the tank is a critical process that requires station and vehicle tank parameters.

- The Type IV tank has been subjected to severe cyclic loading conditions during the fast refueling of compressed hydrogen which affects the stability of the tank. However, Type IV has shown good structural stability during the mechanical and thermal loading conditions.

FUTURE DIRECTIONS

- The structure of the Type IV tank can be modified based on internal loadings.
- The composite tank can be tested for different fiber-based materials or combinations which lower the weight of the tank.
- The communication-based refueling strategy for existing hydrogen refueling stations for better refueling performance
- Incorporation of flow metering or flow regulation in refueling stations can optimize the refueling process.
- Transient heat transfer studies would be advantageous for the exact estimation of heat flow in or out of the tank.

Index

Printed in the United States
by Baker & Taylor Publisher Services